Science

Is An Action Word

Science
Is An Action Word

Peggy K. Perdue

Illustrated by Karen Waiksnis DiSorbo
Foreword by Diane A. Vaszily

Scott, Foresman and Company
Glenview, Illinois London

Good Year Books
are available for preschool through grade 12
and for every basic curriculum subject plus many enrichment areas.
For more Good Year Books, contact your local bookseller or
educational dealer. For a complete catalog with information about
other Good Year Books, please write:

Good Year Books
Scott, Foresman and Company
1900 East Lake Avenue
Glenview, Illinois 60025

To Katy, Jenny, and all elementary students.
May science always be an active part of your school day.

Printed in the United States of America.

2 3 4 5 6 EBI 95 94 93 92 91

ISBN 0-673-38968-5

Contents

Foreword

Elementary students represent the most fertile ground possible for teaching problem solving and for developing enthusiasm for science. But then, science is not just a school subject; it is life itself, the life that is all around us. As teachers, it is our responsibility to open a new view, to share with our students exciting knowledge that intrigues and fascinates them. If you—as the teacher—display awe, wonder, excitement, and enthusiasm, your students will relate, pick up on, and respond to those feelings. If you use an inquiry approach with them, they will follow the same inquiry technique on their own. Your approach to teaching science can be the beginning of their learning problem-solving skills.

The most important ingredient in this approach is not the book, the supplement, or the ditto; it is *you,* the teacher. A room full of apparatus cannot evoke a child's enthusiasm; only what you do with that apparatus can create excitement. Your interest in the activity, your responsiveness, and the enthusiasm you project are the most precious items you can give to your class.

These exciting ready-to-use science activities will keep science "active" in your classroom. Utilizing inexpensive, easily available materials, you can turn your classroom into a laboratory filled with hands-on excitement.

To begin, divide your class into teams of colleagues who will collaborate as they carry out an investigation. Encourage exchanges of ideas within and among teams. Always make the data that each team collects available to the entire class for further analysis, graphing, and drawing conclusions. While the science activity itself certainly has value, it is the investigative process that is most important.

Peggy Perdue designed and used these activities in her laboratory classroom during her years as an elementary school science specialist, and she knows how much students enjoy them and benefit from the process skills they develop. It was always a thrill for me to walk into her classes and see exciting science in action!

Now, in the pages that follow, you can provide your students with that same excitement through these activities. You are about to enter a real science lab—wherever you are!

Diane A. Vaszily
The Science Eye

A Scientific Approach to the World

We confront problems nearly every day of our lives. One of the most valuable skills we can teach our children, therefore, is how to approach the solving of problems. The scientific approach is not limited to solving science problems. It is applicable to all areas of life. The basic steps of the scientific approach are outlined below.

Purpose. When you ask, "Why am I doing this?" you are defining a purpose. With children, the purpose is best stated in the form of a question—e.g., What effect does air pressure have on a metal can? How does water affect rocks? What color candy appears most often in a bag of M&Ms? Such questions give direction to the activity and alert children as to what they should be investigating.

Hypothesis. Children love big words, and "hypothesis" is a big word that means an educated guess. An hypothesis is a guess that answers the purpose—e.g., Air pressure will cause the can to explode. Water makes rocks smooth. There are more red M&Ms in the bag (at least I hope so!).

Because it is a guess, an hypothesis may be right or wrong. Children often become concerned when they discover that their hypothesis is wrong, but scientists often make a wrong hypothesis. Emphasize to the class that it is OK to make a guess that turns out to be wrong. What is important is for the children to make an educated guess about how they think the activity will turn out.

Materials. Ask your students what items they will need in order to do the activity. Then try doing the activity with just those items they listed. Is anything missing? What would they add? At first you will find that the children leave out several essential items, but with continual practice they refine this skill. As students become proficient, try blocking out the materials list on the Student Lab Sheet before duplicating, and then ask the class to fill in the section on their own.

Procedure. The procedure is a set of directions that tells how to do the activity. The procedure helps children by clarifying and sequencing the activity step by step. Once again, as students become proficient at following procedures, you can try blocking out the steps before duplicating the sheets. Eliminating the listed directions leaves the activity open ended for students to complete. Because the various groups will probably perform the investigation in different ways, this technique makes comparisons more difficult, but it does develop problem-solving skills.

Observations. When we think of observing, we think only of what we see. Observing, however, involves all of our senses: seeing, smelling, hearing, touching, and tasting. The more senses you can get the children to use, the more they will understand and remember the activity.

Allow enough time for observations and for discussing what the children are observing. Compare what each child observes. Are all the observations the same? Repeating an activity often increases the detail of the children's observations.

Always encourage students to write down their observations. By recording what they observe, they will have something to refer to later that will help them remember the experience. Such records may take the form of charts, graphs, drawings, or photographs. When children have the opportunity to record their observations in some way, their observations become important and valuable—a great boon to the development of their self-esteem.

Conclusion. The conclusion is probably the most difficult part of the scientific approach. Many students merely restate what they have observed and call that their conclusion. Explain that the conclusion should explain the results. Encourage them to use the word "because." For example: The rocks became smooth because they rubbed against each other. While a conclusion may not be thorough, it should pull all the component parts of the activity together into a meaningful statement for the students, a statement they can relate to their lives.

How to Use This Book

Even if you've never taught science before, **Science Is An Action Word** will quickly make you feel comfortable and excited about sharing science with your students. The book covers activities in four areas: scientific method, earth science, life science, and physical science.

Place your students into lab teams whenever possible. Not only does teamwork allow for a sharing of ideas, but it also helps reinforce the concepts presented. Emphasize that scientists rarely work alone, and encourage students to discuss their observations.

Each activity is composed of two parts: an explanatory section for the teacher and a Student Lab Sheet. The teacher explanatory section begins with a brief introduction designed to give an overview of the activity's main concept. The introductions should help you decide at a glance the sequence of labs you would like to do with your class.

Following the introduction, the teacher explanatory section lists the materials you will need for lab teams ("Materials per Lab Team") and for individual students ("Materials per Student"). Activities that involve a class demonstration or special items for the entire class also have a list of "Materials per Class."

Most of the materials required for the labs in **Science Is An Action Word** are already in your classroom. If you are lacking a few items, make a list, duplicate it, and give a copy to each student. You can tell the class to go on a treasure hunt for the items on the list, but be sure to go over the list with the students before they go treasure hunting! Generally, children are delighted to contribute to their classroom science lab.

How many times have you started an activity only to find that you forgot to arrange for a movie projector, duplicate a worksheet, or make an essential visual? To avoid such frustrating experiences, **Science Is An Action Word** outlines the necessary advance preparation under the heading "Preparation." Under the following heading—"Focusing Activity"—you will find information about any special skill or knowledge that students need (e.g., being familiar with terms such as "fulcrum," "lever," and "work").

Under the heading "Procedure for Lab," you will find detailed instructions on how to do the activity. Included in this section are questions to ask and even some of the possible answers to anticipate from your class. Review this section prior to class, and familiarize yourself with the material it contains. While reviewing it, ask yourself the suggested questions. Also ask yourself what other questions you should pose to the class.

Finally, read through the section under the heading "Extension Activities." Here you will find ideas on how to expand the lab activity into an entire unit and how to integrate the activity with other areas of science.

The Student Lab Sheets are designed to be self-explanatory. Most follow a standard format with the following headings: "Purpose," "Hypothesis," "Materials," "Procedures," "Observations," and "Conclusion." These headings, of course, correspond to the major components of the scientific approach to problem solving.

Science Teaching Tips

Science is not an isolated subject. Therefore, in addition to the block of time you set aside for science instruction, try to combine science with other subjects. You may also wish to set up a learning center where students can explore items in their world. The learning center may be just a corner of your room, an area that can accommodate no more than a small table or a TV tray.

Look around your classroom for counter space or a shelf that you can clear and make into your "science spot." A science spot is a place where students can work independently on science activities. It may be as simple or elaborate as you wish. Just keep in mind that students should be able to carry out an activity there with minimal guidance.

Consult your school librarian about science resource books you can include in your reading corner. Having such books easily accessible allows students to research topics of interest and perhaps look up answers to questions they may have. Be sure to include science resource books at a variety of reading levels in your reading corner.

SCIENTIFIC METHOD

LAB EXPERIENCES

You're the Apple of My Eye

This lab is a great one to sharpen students' observational skills. You can also use it when teaching a unit about nutrition and when developing student classification skills.

Materials per Student

One small apple
Paper towel
Paper lunch bag
Set of crayons
Plastic knife
Student Lab Sheet 1.1
Pencil

Preparation

Duplicate Student Lab Sheet 1.1.

Focusing Activity

Explain that scientists must be able to make good observations. To make good observations, scientists rely on their five senses to give them accurate information.

Procedure for Lab

Hand out one apple and one paper lunch bag to each student. *Tell them not to eat the apple* (get this instruction in early!). Remind the class that everyone is a scientist and that they should all observe the outsides of their apples to get as much information as possible. Forewarn the class that in a few minutes you will be asking them some questions to see how well they observed their apples.

After a few minutes, tell the students to place their apples in the lunch bags and then to fold the tops of the bags down. Give each student a copy of Student Lab Sheet 1.1. Explain that they are to draw a picture of their apple on the lab sheet in the section marked "How I Remember My Apple." While they are drawing, ask the students questions that pertain to their observations. Some examples might include: Is your apple more round or more oval? Is it hard or soft? What color is your apple? Is it the same color everywhere? Could you wrap your hands all the way around your apple? Is your apple's skin smooth all over? Does your apple have a stem? Could you smell your apple? What did it smell like?

After students complete their drawings, have them remove their apples from the bags. Were their observations accurate? What would they change in their drawings now that they can see their apples once again? Direct students to the section of Student Lab Sheet 1.1 labeled "What My Apple Really Looks Like," and have them draw their apple while looking at it.

Ask the students whether the inside of their apple looks like the outside. Ask how they could find out, and allow enough time for them to hypothesize possible solutions. Then have each student use a plastic knife to cut the apple vertically. Since even plastic knives can cut fingers, however, be sure to review the proper way to handle a knife before proceeding with this portion of the lab.

Encourage the students to make observations as they cut open their apples. If you have extra apples, cut several horizontally. Ask students to note any differences in the appearance of apples depending on whether the cuts are made vertically or horizontally.

Next, have the students remove the seeds from their apples, noting how many seeds each apple has. Do all apples have the same number of seeds? After they remove and count the seeds, students may taste their apples. This is a good time, incidentally, to remind students never to eat anything without first obtaining an adult's permission. Ask the students what observations they can make about the taste of their apples. Ask them in which food group the apple belongs.

Extension Activities

1. Have students graph how many seeds their apples have. The easiest way to do this is for students to prepare bar graphs on sheets of the reproducible graph paper found at the back of this book. Start by having the children label the axes and then put one seed in each square going up. Once they have the seeds lined up, have them remove the bottom seed in the column and color the square underneath. Have them continue removing and coloring until the entire column is colored in and no more seeds remain on the graph paper.
2. To develop student classification skills, have several varieties of apples available in the classroom. Ask the students to sort the apples by color and/or size.
3. If possible, take the students on a field trip to a local apple orchard.
4. Use the apples in simple math story problems. For example: If Katy has 16 apples and gives 6 to Jenny, how many apples does Katy have left?
5. Have an apple eating day in which students bring different recipes that include apples as one of the main ingredients. Choose a simple recipe to prepare in class. Following a recipe gives students good practice in measuring and sequencing.
6. Have a tasting party. Ask for volunteers to bring in various foods made from apples. Serve apple juice to drink.
7. Do apple printing. Start by placing a small amount of tempera paint in a shallow dish. Then have students dip a halved apple into the paint and press onto paper. Ask them to compare prints made from apples cut vertically with those made from apples cut horizontally.

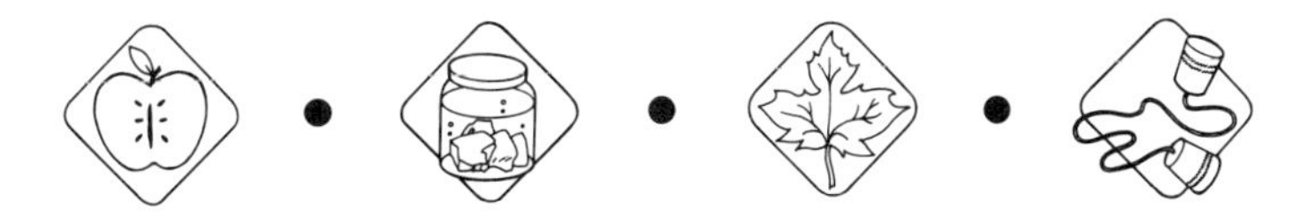

 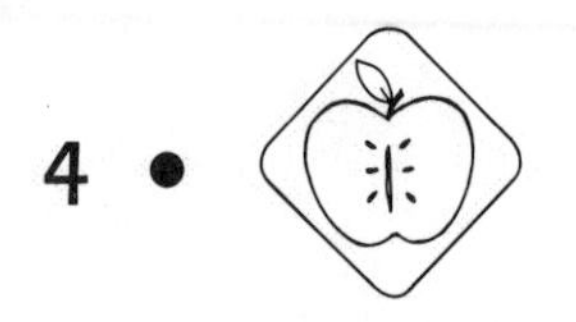

Here's Looking at You!

When doing lab experiments, students often confuse observations with inferences. This lab will help them understand the need to be objective when conducting science experiments.

Materials per Lab Team

Paper towel
Instant oatmeal
Sheet of paper for recording observations
Pencils

Preparation

Prior to the lab (i.e., when students are not around), place a small amount of instant oatmeal on each paper towel (one per lab team).

Focusing Activity

Discuss the terms "observation" and "inference." Ask how the words are different in meaning. Students should understand that an observation is information gathered by the five senses while an inference is a conclusion drawn from the observation.

Now is also a good time to discuss different ways to record observations. Students should be familiar with charts, graphs, drawings, pictures, and other visual means of recording observations as well as written notes of what they observed.

Procedure for Lab

Hand out the "mystery substance." Tell students that they should write down as many observations about the mystery substance as they can. Encourage them to use their senses, but decide in advance whether you are going to allow students to taste the substance. If you allow them to taste it, emphasize that they must never taste anything without first receiving permission from an adult.

Circulate among the lab groups, encouraging each to perform "mini experiments" that will enable them to get as many observations as possible. For example, you might inquire, "Does it dissolve?" or "How would you describe the size of the particles?" The latter question is particularly good because students are likely to say that the particles are small. Point out that size is relative—i.e., the particles are small compared to a pencil but large compared to the head of a pin. Encourage students to be exact in describing their observations.

After students have observed for 10 to 15 minutes, discuss their lists. Are all the items they wrote down observations? Ask the students how they could make their observations more exact and less subjective.

Finally, have the students write down what they can infer the mystery substance is based on their observations. Wait until everyone has had an opportunity to make an inference before revealing the answer.

Extension Activities

1. Try the same lab with another substance. Repeating an experiment tends to sharpen observational skills.
2. Cut out a full-page picture from a magazine. Glue the picture onto one interior side of a file folder. Then cut little windows in the other side of the folder so that, when the folder is closed, little bits of the picture become visible as you open the windows. Hold up the closed folder in front of the students and open one of the windows. Ask what observations the students can make. Open another window, and ask whether they can infer what the entire picture is.

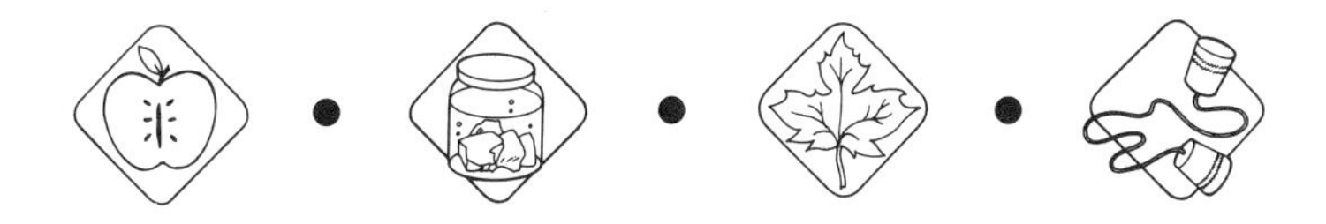

Do You See What I See?

Observational skills are important in science. This lab involves students in observing properties of matter and recording what they find.

Materials per Lab Team

Macaroni, 6 pieces
Testing materials (e.g., magnets, magnifying glasses, paper towels, rulers, water, and clear plastic drinking glasses)
Piece of paper for recording observations
Pencils

Preparation

None

Focusing Activity

Ask students how they would describe a peanut butter and jelly sandwich to you if you had never seen one before and wanted to know exactly what it was. Give each student an opportunity to respond, and write all responses on the board. Go over the list with the class, determining which statements help the most in understanding the exact nature of the sandwich. Explain that an object's size, color, shape, texture, and weight are just a few of its *properties.* Discuss how some properties can be very obvious while others are revealed only through experiments or investigations.

Procedure for Lab

Give each lab team six pieces of macaroni. Tell the students that they are to write down every property they can discover about the objects you just gave them. Explain that they must be able to prove to the class that the objects do indeed possess the properties listed. For example, they must not write down that the objects could be used in a macaroni and cheese dinner because they have not performed the experiment in class to test for that property. Inferences do not count!

Allow time for students to investigate. Provide some additional investigative tools—e.g., magnets, magnifying glasses, rulers, water, clear plastic drinking glasses, and other materials—at a location where students can have access to them. If necessary, ask questions to encourage a thorough investigation. What happens if they blow on the objects? What happens if the objects come in contact with water? How flexible are they? Are they magnetic? Encourage students to use all of their senses in the investigation—with the possible exception of their sense of taste.

When the lab teams have finished compiling their lists of properties, create one large class list. Doing so may prompt students to think of even more properties or of different investigations to discover additional properties.

Extension Activities

1. Place a mystery substance at the learning center. Have students list four of its properties.
2. Have students repeat the macaroni experiment using a lima bean instead of the pasta. What properties do the macaroni and lima beans share in common? What properties does each possess that the other lacks?

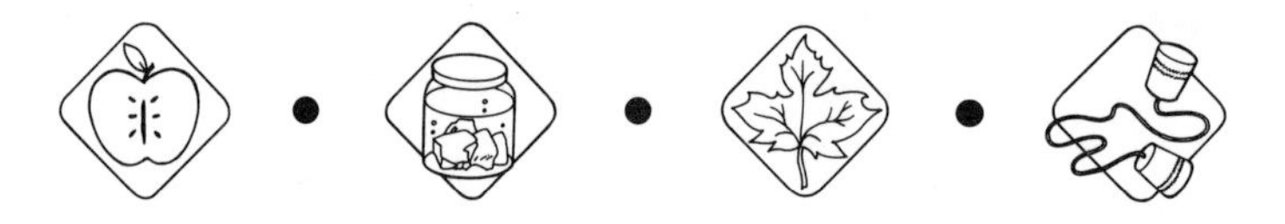

A Bagful of Color

This is a great lab for teaching both science skills *and* counting, sorting by color, and one-to-one ratios; it also can help develop fine motor skills. Students will make an hypothesis, graph observations, and have fun—all at the same time!

Materials per Student

Snack-size package of M&M's®
Crayons: green, yellow, orange, brown, red, and black
Student Lab Sheet 1.2
Pencil

Preparation

Duplicate Student Lab Sheet 1.2. Make sure students understand one-to-one relationships.

Focusing Activity

This lab is a good introduction to making an hypothesis and graphing observations.

Hold up a bag of M&M® candies and ask students what they think is inside. When they answer "M&M's®," ask them to be more specific. Ask them to name the colors of the candies. Write the color names on the board as students respond: red, yellow, green, orange, light brown, and dark brown.

Now ask the students how many candies of each color they will find in the bag. Some students will try to give you an exact answer. Ask these students how they know the exact number of each color. It won't be long before students tell you to open the bag and let them count!

Before opening the bag, however, have the students write on the Student Lab Sheet which color they think will appear most often. Tell them that by putting down a number they are making an hypothesis, or an educated guess. *Emphasize that a guess may be right or wrong, and reassure them that the correctness of the guess does not matter.* Some students become upset when their hypothesis turns out to be incorrect, and they try to change it. Tell such students that even great scientists often make an incorrect hypothesis.

Procedure for Lab

Hand out the M&M's® packages. Remind the students not to start the lab until you give the OK.

Take some time explaining the graph paper, especially the x-axis and y-axis. Point out how the numbers go up on the y-axis. Explain that they must first sort out the M&M's® by color and then graph their results. For example, ask them how many squares they would color orange if they find five orange M&M's® in the bag. The easiest way to have students understand the graphing concept is to have them actually place the M&M's® on the graph paper (one per square) in the appropriate color columns. This technique also allows you to see any difficulties students may be having before they complete their graphs.

Have the students open their bags, sort the candies, and lay the M&M's® on their graph paper. If anyone finds more than 11 candies of the same color, ask the class to create a way to show the result. You might suggest one solution: change the numbers on the y-axis so that they increase by 2s.

After the students have their candies correctly positioned on the graph paper, instruct them how to color the graph. Tell them to *start at the bottom of the first column,* remove one M&M®, eat it, and color the square below it the same color as the M&M®. Indicate that they should repeat this procedure, column by column, until they have removed and eaten all the M&M's® and correctly colored in the graph paper squares.

By looking at the colored columns, students should be able to tell which color candy appeared most often in their bags of M&M's® and which color appeared least often. Explain that graphs are very useful in showing us results at a glance. Then tell them to transfer the data from their graphs to the chart that lists the totals for each color.

Finally, have the students check their findings against their hypothesis. Mention again that there is nothing wrong about making an hypothesis that turns out to be incorrect. Encourage students to speculate on why a particular color appeared most often in the bags of M&M's®.

Extension Activities

1. Combine all the individual results into a single graph for the entire class. Which color M&M® has the highest number?
2. Have the class write a letter to the Mars Company, inquiring how it chooses the quantities of each color to put into bags of M&M's® and whether it attempts to put an equal number of each color into the bags.
3. Repeat this experiment with a mixed-fruit pack of Life Savers®.
4. If students understand the difference between graph formats, have them redo their bar graphs as line graphs. Make sure that they place the numbers on the y-axis even with the horizontal lines (not between them as shown on the Student Lab Sheet). Similarly, have them match up the color names along the x-axis directly with the vertical lines (not between them as shown on the Student Lab Sheet).

   For example, a student who has five green M&M's® would follow the vertical line upward from the word "GREEN" until reaching the fifth horizontal line up from the x-axis. There he would place a dot at the intersection of the two lines. After repeating this procedure for each color, the student would then connect the dots as in a dot-to-dot puzzle.

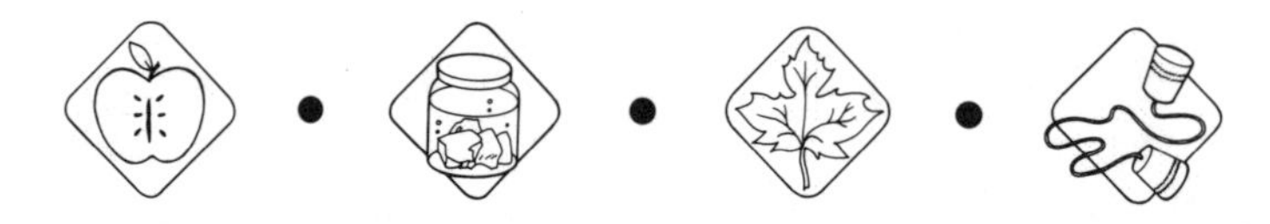

Classified Information

Scientists must be able to classify accurately. We classify rocks, minerals, plants, animals, and more. Students may be able to describe qualities that make a robin a bird or that make a snake a reptile, but do they understand the complexity of the scientific classification system? After this lab, their classification skills will be more refined and their appreciation of the classification systems that scientists use will be greater.

Materials per Lab Team

Miscellaneous office supplies (e.g., rubber band, pencil, paper clip, brad, eraser, staple, piece of tape, sheet of memo paper, tack, ruler)
Small tray or paper plate for carrying supplies

Preparation

Gather the office supplies. Make up a complete set for each lab team, or—depending on the maturity level of the students—allow teams to take responsibility for assembling their own supplies from a central location.

Focusing Activity

Ask students how we know that a robin is a bird. Students should reply that a robin's characteristics—has feathers, lays eggs, and is warm-blooded—make it a member of the bird (Aves) class. Accept all responses. Then ask the students how they think the animal classification came about.

Procedure for Lab

After each team has its set of office supplies, explain that each team must classify the "unknown objects." Have students work as teams to make up logical groups. Explain that you will be coming around to challenge each team's classification system and that they must be prepared to defend their groupings.

Allow teams a few minutes to get started. Then begin playing the devil's advocate. You need find only one flaw to challenge a system and send students frantically rearranging their groups. You will be amazed at the creativity that students display in trying to make their classification systems foolproof.

The problem that most frequently afflicts classification systems is the use of just one characteristic as a basis for grouping. Remind students that scientists use several characteristics when defining classes of the animal kingdom. Once students understand this principle, you may find it exceedingly difficult to find flaws in their systems. On the other hand, if they go to the opposite extreme of including too many characteristics, each of their groups will consist of the one member that meets all the criteria—not a very practical system of classification.

Discuss with the teams how they felt during the lab. Ask whether they understand why you could find flaws in their systems and what they would do differently the next time. Have the teams compare classification systems, noting similarities and differences.

Extension Activities

1. Create a bulletin board displaying members of the animal kingdom. Feature a mystery creature each day. Present a picture and a brief paragraph describing the creature, and have students guess the class to which it belongs.
2. Repeat the lab using different materials. Compare before and after strategies.
3. Plan a trip to the zoo. Have students chart the various animals by class there, and have them determine which class has the most members at the zoo.

Student Lab Sheets

SCIENTIFIC METHOD

Name________________________ Date________________

• You're the Apple of My Eye

How I Remember My Apple

What My Apple Really Looks Like

Cross Section of Apple

Other Observations

Number of seeds____

Do other apples have the same number?________________________

Taste____

Food group____

Name________________________ Date________________

A Bagful of Color

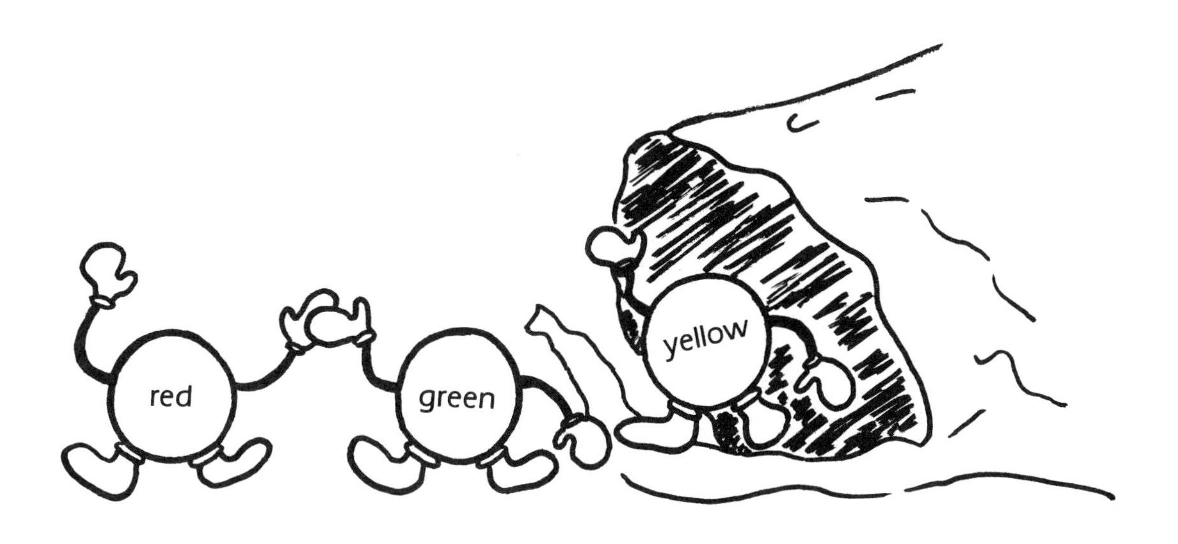

Purpose

Which candy color will appear most often in a bag of M&M's®?

Hypothesis

____

____

____

____

Materials

Snack-size package of M&M's®
Crayons: green, yellow, orange, brown, red, and black

Observations

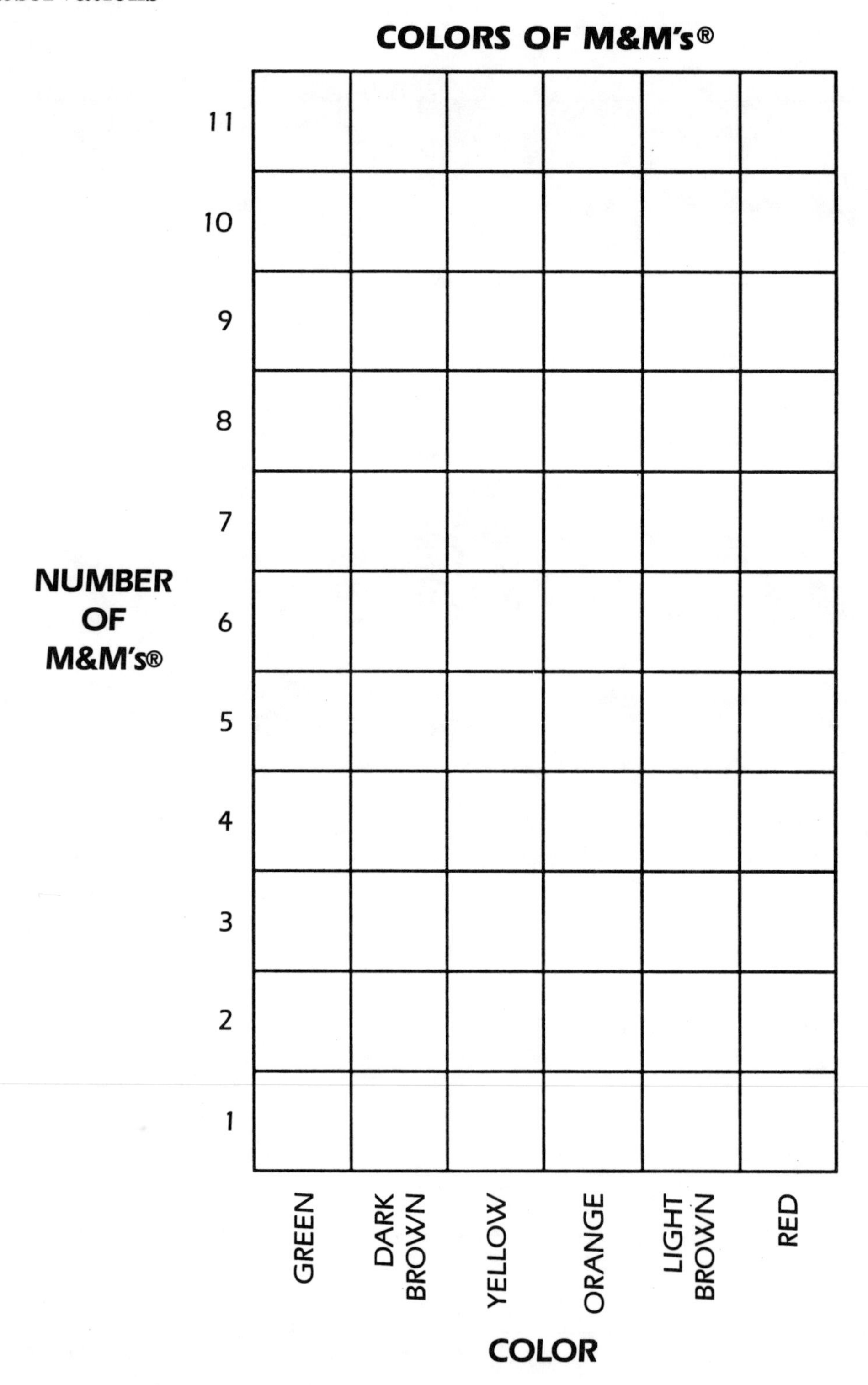

Transfer the data from your graph onto the chart below.

COLOR	Green	Dark Brown	Yellow	Orange	Light Brown	Red
NUMBER						

LAB EXPERIENCES

Fun-ominal Fossils

Children in the primary grades love dinosaurs. Not only can they memorize the long names of the giant lizards, but they can also readily identify models of the different species. How do you capture this enthusiasm and direct it into a hands-on lab experience? Have students create and study their own fossils!

Materials per Student

Disposable dish (aluminum, paper, or plastic)
Lab shirt
Plaster of Paris
Bone (from chicken, turkey, or ham)
Tongue depressor or craft stick
Newspaper
Petroleum jelly
Water in a small paper cup
Water in a plastic drinking glass

Preparation

Check with a local delicatessen to see if they will give you (or sell at a nominal charge) 1/4-pound containers for use in this lab. Paper bowls and aluminum pie tins also work well.

Explain to your class that each child must bring in a small bone. Show the children the dish they will be using in the lab; the bone should be small enough to fit inside the dish (chicken, turkey, and ham bones work best). Students should thoroughly clean the bones *before* bringing them to school (expect a few children to miss this part of the instructions!) by first boiling the bones to remove excess meat and fat and then allowing the bones to dry. In addition, you may wish to soak the bones in a ten percent bleach solution for a few hours and then put them out in the sun to dry.

Focusing Activity

Introduce the activity with a small example of creative dramatics. Tell the students that you would like to show a friend of yours a live dinosaur. Explain that your friend (invent a name) wants to pet a dinosaur but that you have a problem: You don't know where to find one. Where can a person go these days to pet a dinosaur?

Students may respond: "Go to the museum." Ask if the museum has live dinosaurs. Students will quickly tell you that dinosaurs aren't alive anymore. Ask if anyone in the class has ever seen a live dinosaur. When they answer no, ask the students how they know dinosaurs ever existed. Some students may offer pictures as proof, but you can reply that there were no cameras or video recorders when dinosaurs roamed the Earth. As a matter of fact, no humans (not even Fred Flintstone) were around in the dinosaur age.

Again ask how we know that dinosaurs existed. Most likely, at least one student will answer: "Museums display dinosaur remains." At this point, tell the class that they are about to become archaeologists in search of animal remains.

Procedure for Lab

Have each student put on a lab shirt (a men's shirt worn backwards is perfect) to protect his or her clothing. Cover the lab surface with newspaper to protect it.

Have students coat their bones with a *thin* layer of petroleum jelly. While they are doing this procedure, fill a plastic cup one quarter full of water for each student. Then have the students add plaster of Paris to the water, a spoonful at a time, stirring the mixture with a craft stick after each spoonful. Have the students keep adding plaster of Paris and stirring until the mixture is smooth and the consistency of soft-serve ice cream.

At this point, students should empty the mixture into a disposable dish and place their bone on top of the mixture, pressing down on it *slightly.* If they submerge the bone completely in the plaster of Paris, they will find it difficult to remove later on. Then allow the plaster to dry; depending on the consistency of the mixture, the drying process can take from an hour to overnight.

Ask students to describe the ground after a hard rain. Relate their description to the plaster of Paris. Ask what would happen if they were to go for a walk across the rain-soaked ground (besides possibly ruining their shoes). Explain that when a dinosaur died, it fell on the soft ground and left a "print" that was exactly like its body. Soon nearly all of the dinosaur's body decayed and disappeared so that only the bones remained. Over the years, these bones were covered by dirt and mud until they could not be seen. Thousands of years later, archaeologists dug deep into the ground and discovered the dinosaur remains. From these remains, archaeologists were able to gain a very good idea of what different dinosaurs looked like.

After the plaster of Paris has dried, have students carefully remove their bones and examine their "fossils." If magnifying glasses are available, let them take a closer look. What can they see? Does the fossil look like the bone? Review how they made their fossil and how fossils are made in nature.

Extension Activities

1. Have students match bones to fossils at a learning center.
2. Create fossils from several other items in nature (feathers, shells, leaves, rocks, etc.). Display the fossils at a learning center and have students try to identify them.
3. Go on a fossil-hunting expedition around the schoolyard. See whether students can find any fossils and, if so, whether they can identify what made them.

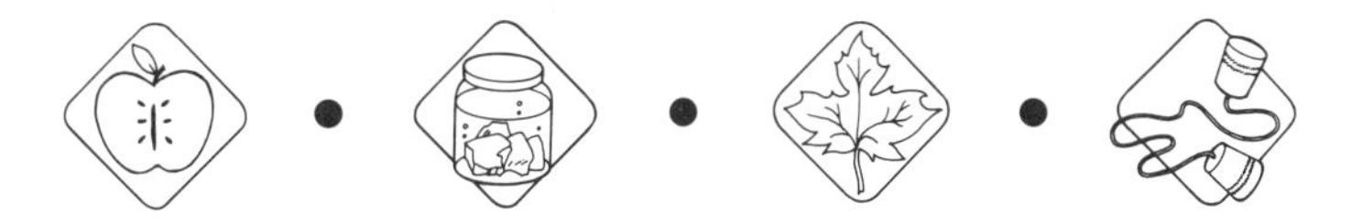

Wearing Down

What happens when water comes in contact with rocks? Erosion. In this lab, students will simulate nature as they combine water and rocks in a mini-environment.

Materials per Lab Team

Paper towel
Plastic peanut butter jar with lid, empty
Small sharp stones, 10 (a "soft" rock, such as limestone, will provide the most dramatic results)
Plastic drinking glass
Masking tape

Materials per Student

Student Lab Sheet 2.1
Pencil

Preparation

Collect empty plastic peanut butter jars. Wash and remove labels. Duplicate Student Lab Sheet 2.1.

Focusing Activity

This lab provides a good introduction to erosion. Students need not have any background knowledge of the subject.

Divide the class into lab teams of two students each. Assign each team a letter, either A or B. It is important for teams to remember their assigned letters because the letter determines which lab procedure they will follow.

Ask students to think of any place where water comes in contact with rock. This will help focus their thoughts. Accept all responses.

Procedure for Lab

Refer students to the lab sheet. Read the purpose together. Have students write their hypothesis on the lines provided. You may wish to remind students that a hypothesis is an educated guess that may be proven right or wrong at the conclusion of the lab. A mistaken hypothesis is perfectly all right. In fact, a good hypothesis is not necessarily a correct hypothesis.

Explain that each team must read the procedures very carefully. Since different teams will be doing different parts of the experiment, it is essential that they remember their lab letter.

As a control for this experiment, set up a jar that is half full of water. Put ten stones inside the jar and screw the lid on tightly. Allow this jar to sit undisturbed while students complete the experiment. Ask students why it is important to have a control in an experiment. Make sure they understand that the control allows variables to be isolated so that valid comparisons can be made.

Permit students to proceed through the experiment on their own while you circulate among the lab teams. Ask students what they think they will find when they stop shaking the jar. Will A teams discover something different from B teams? What will the difference(s) be?

When everyone is finished, combine an A team with a B team so that the two teams can compare results. Students should then fill in the lab sheets charts, making detailed observations.

Discuss the changes they observed. They should understand what made the difference: Water combined with shaking caused the most erosion. Ask where similar situations occur in nature. The correct answer is wherever moving water—a river, stream, or ocean—comes in contact with rocks. If students merely answer "in water," remind them that the stones in water in the control jar remained unchanged.

As students think about the changes they observed, they should write their conclusions. They should try to explain why the water changed the rocks, noting that more erosion took place in the jar containing rocks and water than in the jar with just rocks.

Make sure that students know what we call the process by which water wears away rocks. That process is erosion.

Extension Activities

1. Have students repeat the experiment, this time shaking the jars even more. Does more shaking change the results?
2. Have students look around the school property for signs of erosion. If they find any, ask them to figure out what caused the erosion and what might be done to stop it.
3. Have students look around their homes and community for signs of erosion. Ask them to write a brief report about what they find.
4. Invite a landscaper to speak to the class about what can be done to prevent erosion.

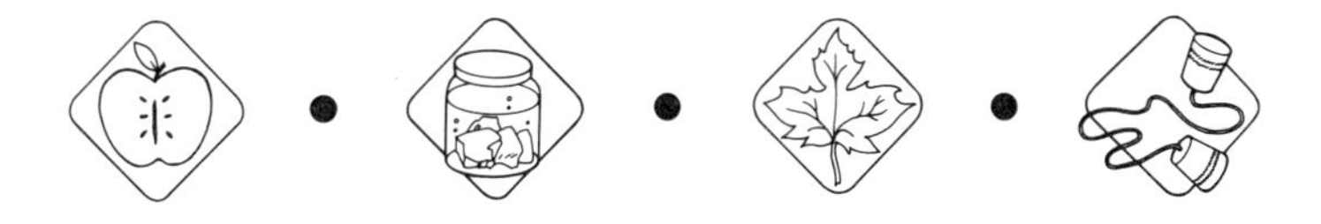

Clean as a Mountain Stream

Many industrial companies seem to think that the chemicals they dump into our country's rivers and streams are of little consequence. After all, the world has lots of water, right? In this lab, students will observe firsthand that even a small amount of pollution is very difficult to "wash" from the water. This experiment is intended to raise students' consciousness about environmental pollution.

Materials per Lab Team

Large mayonnaise jar, empty
Plastic drinking glasses, 10
Water
Food coloring (red, blue, or green)
Marking pencil
Sponge

Materials per Student

Student Lab Sheet 2.2
Pencil

Materials per Class

The Lorax by Dr. Seuss

Preparation

Practice reading *The Lorax* so that you can read it aloud with feeling. Duplicate Student Lab Sheet 2.2.

Focusing Activity

Read Dr. Seuss's *The Lorax* to the class, and then discuss the book with your students. How do they feel about the story? Could anything like that actually take place? Why or why not?

Ask the students what actual evidence they have seen that pollution exists in our society. Accept all answers. If no one has seen any, take the class on a quick trip around the schoolyard in search of signs of pollution. Discuss why everyone should be concerned about the amount of pollution in our environment.

Procedure for Lab

Refer students to the purpose of the lab as stated on their lab sheets. Before students fill in their hypothesis, however, read the procedures together. Then tell the students to predict how much water they will need to add in order to make the food coloring disappear. Ask the students to predict which glass of water will no longer show the color change, and tell them to consider this information as they write their hypothesis.

The more glasses of water required to dilute the coloring, the more difficult it would be to remove the "poison" from the water. Explain that while the lab sheet refers to food coloring as poison, this is just for the sake of the experiment; in actuality, food coloring is not poisonous.

Each lab team should have one student gather the necessary materials. Then the lab team should perform the experiment. When all lab teams are finished, discuss the results as a class. If the water in the glasses became clear, is this proof that the "poison" was no longer present? Why or why not? Students should understand that some chemicals dissolve in water and cannot be detected by sight alone.

Ask the class how this lab relates to the "real world." Encourage students to brainstorm, and allow time for a discussion in which students can express their feelings about environmental pollution.

Extension Activities

1. Take a field trip to a waste water treatment plant.
2. Put up a bulletin board devoted to environmental topics. Have students bring in newspaper clippings, photographs, or other visual evidence documenting pollution.
3. Work with the class to create a play about a city that allows no pollution. Perform the play for other classes in the school.
4. Start an antipollution campaign at your school. Have the students design materials for the campaign and have them put it into action.

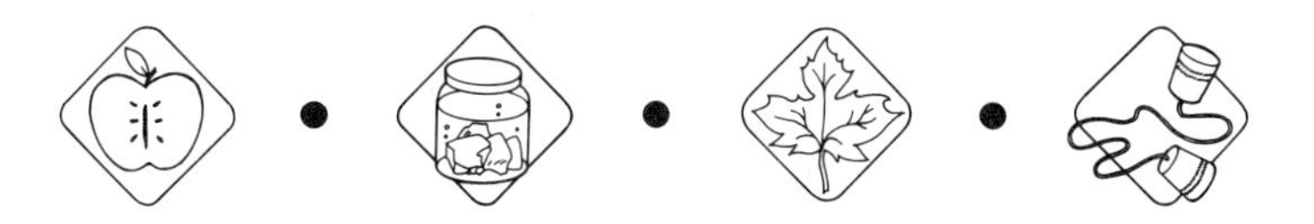

Layer Cake

Students often find rocks that fascinate them. They wonder how such rocks form and acquire their unique shape, smoothness, and/or color. In this lab, students will observe the layering process that occurs in the formation of sedimentary rock.

Materials per Lab Team

Plastic peanut butter jar, empty
Small rocks
Potting soil
35mm film container, empty
Sand
Water
Clock

Materials per Student

Student Lab Sheet 2.3
Pencil

Preparation

Collect enough plastic peanut butter jars so that each lab team has one. The jars need not be identical. In fact, if you have a variety of jars, you could ask students to compare results to see whether the container makes a difference in the experiment.

Contact a film-processing company for the 35mm film containers needed for this lab. Quite frequently, such companies will respond to a request by collecting containers for you over a period of time. Duplicate Student Lab Sheet 2.3.

Focusing Activity

Students should be familiar with types of rocks: sedimentary, igneous, and metamorphic. For review purposes, ask how rocks are formed. Accept all answers (sometimes difficult to do when students respond incorrectly after you've spent several lessons on the concept!). Ask if all rocks are formed in the same way. Explain that in this lab they will be observing the beginnings of sedimentary rock– i.e., rocks formed from sediment.

Procedure for Lab

After students indicate that they know the meaning of sedimentary rock, refer them to their lab sheet. Read the purpose of the lab together. Then remind them that an hypothesis is an educated guess that might be proven right or wrong. Have each student (or team) write an hypothesis.

Have the students read over the list of materials and the procedures for the lab. When they finish, ask if they have any questions. Once you've answered their questions, have each lab team select one person to gather the needed materials. The teams may begin the experiment as soon as they have their materials.

Students should make their first observation while the contents of the jar are still in turmoil. Encourage students to pay attention to detail and to record what they see by drawing on the lab sheet.

While waiting for the sediment to settle, ask the class several questions. What did the water look like when you were shaking it? (All mixed up.) Where in nature could you find water and sediment getting mixed up? (In a swift-moving river or stream and along the edge of the ocean, especially during a storm.) What appears to be happening as the water settles in the jar? (The larger, heavier pieces fall down first, followed by other sediment.)

When the water has settled, encourage students to observe the jar closely. Ask what they see. Their response should be, "Layers." They should then draw what they see on their lab sheet. Remind students that an observation illustration should accurately represent the experiment.

When everyone has finished the drawing, focus student attention on the conclusion. This is the hardest part of a lab report. Students are tempted to write what they saw happen, merely putting their observation into words. But a conclusion should answer the question, *Why?* To help students write a valid conclusion, encourage them to include the word "because." For example, "The rock fell to the bottom first *because* it was heaviest and the water could not support it as easily as it did the lighter sediment." Or, "The sand could be easily carried by the moving water *because* it was lightest."

After discussing the results of this lab, students should understand the role that sediment weight plays in the layering process of rock formation.

Extension Activities

1. Have students repeat the experiment, this time weighing the particles. Can they prove that weight affects the layering process?
2. Have the students devise a way to measure the depth of each layer in the jar.

Raindrops Keep Falling . . .

April showers bring May flowers, but what brings the showers in the first place? In this lab, students will experience a rain storm and see how evaporation and condensation take place.

Materials per Class

Hot plate
Water
Measuring cup
Small saucepan
Aluminum cake pan, 8 or 9 inches square
Hot pad
Ice

Materials per Student

Student Lab Sheet 2.4
Pencil

Preparation

Duplicate Student Lab Sheet 2.4.

Focusing Activity

Walk to the front of your classroom carrying an umbrella and wearing a raincoat, rain hat, and boots. Keep glancing upward to the "sky." At the front of the room, put down the umbrella and take off the rain gear; don't forget to shake off the water!

Ask your students where rain comes from. Accept all answers without comment.

Review safety procedures, emphasizing that students must be careful around all electrical appliances and especially a hot plate. In addition to reminding them of the heat that a hot plate generates, warn students about the danger of tripping on the electrical cord.

Procedure for Lab

Direct students to their lab sheet. Read the purpose together. Ask students if they believe that you can create a rainstorm right in the classroom. Tell the students that you want them to think of an hypothesis that answers the purpose. If they don't believe that you can create a rainstorm in the classroom, for example, their hypothesis should be that a rainstorm will not form.

Next, direct student attention to the observation section of the lab sheet. Explain how important it is for them to record observations accurately. For this experiment, they are going to draw what they see happen. Turn the hot plate on to its medium-high setting. Put two cups of water into the saucepan and place the pan on the hot plate. While waiting for the water to heat, ask the students what the pan of water represents in

nature. They should respond by saying a lake, ocean, river, or other body of water. Ask how any body of water on Earth is warmed, and they should answer by saying that the sun does the warming. Put ice in the aluminum pan, enough to cover the bottom of the pan. When the water is warm enough that you can see steam rising, ask the students what they see happening to the water. Discuss the meaning of the word "evaporation."

Now move the pan of ice so that is approximately 18 inches above the saucepan and hold it there. You may wish to use the hot pad to hold the aluminum pan because it will have become very cold from the ice. Ask your students what the pan of ice represents in nature; they should answer "clouds."

Allow small groups of students to come up to the demonstration table for a closer look, but remind them of the need for safety. Ask what is forming on the bottom of the aluminum pan. Students should respond "water droplets." It will not take long for these droplets to grow in size. When they do, their weight will make it impossible for them to remain on the underside of the pan. They will begin to fall down as raindrops.

Discuss condensation. Then direct the students to draw on their lab sheets what they have just observed. Clean up the demonstration area while they are completing their lab sheets, and then review the water cycle—both as it occurred in the lab and as it takes place in nature.

Extension Activities

1. How long does it take a cup of water to evaporate in the classroom? Pour one cup of water into an empty plastic peanut butter jar. Use a grease pencil to mark the top of the water. Leave the top off the jar, and measure the water level every three days.
2. At the learning center, place a variety of pictures depicting the water cycle. Have the students put the pictures into the correct sequence. You can make this a self-checking activity by numbering the pictures on the back.
3. Make a temperature graph by coloring in the appropriate squares on a bar graph each day. You can also graph the number of rainy days during the month.

Student Lab Sheets

EARTH SCIENCE

Name________________________________ Date____________________

• Wearing Down

Purpose

How does water affect rocks?

Hypothesis

____

____

____

____

Materials

Paper towel
Plastic peanut butter jar with lid, empty
Small sharp stones, 10
Plastic drinking glass
Masking tape

Procedures for Team A

1. Label your paper towel "A." Place a small piece of masking tape on your plastic drinking glass and label it "A" also.
2. Put ten stones in your jar.
3. Go to "Procedures for Both Teams."

Procedures for Team B

1. Label your paper towel "B." Place a small piece of masking tape on your plastic drinking glass and label it "B" also.
2. Put ten stones in your jar.

3. Fill your jar half full of water
4. Go to "Procedures for Both Teams."

Procedures for Both Teams

1. Screw the lid on the jar tight.
2. Have each person on your team shake the jar 500 times. To give arms a rest, one team member may switch to another after 100 shakes.
3. Unscrew the lid from the jar. Members of Team B should pour the water from the jar into the glass.
4. Empty the stones from the jar onto the paper towel. Look for any changes in the stones.
5. Compare changes in the stones from the Team A jars and the Team B jars. Also compare with the control jar.

Observations

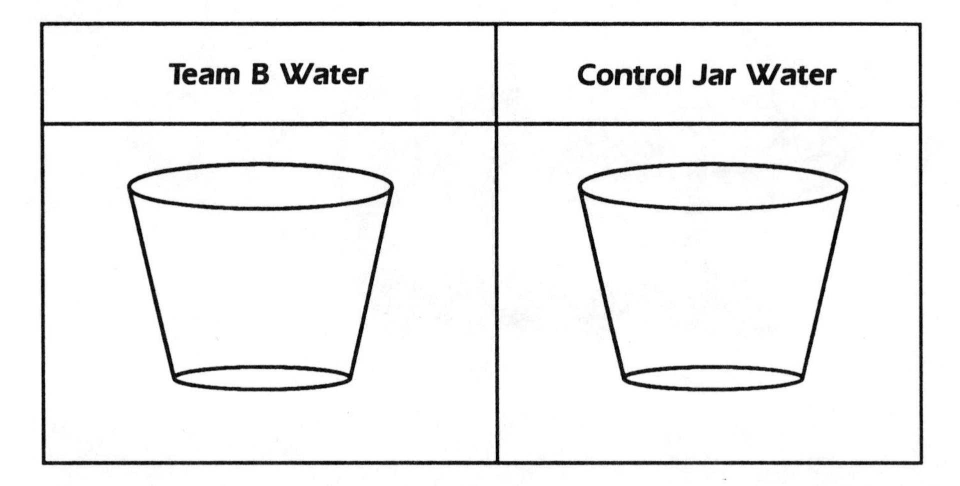

GROUP A STONES	GROUP B STONES	CONTROL JAR STONES

Conclusion

____

____

Name________________________________ Date____________________

Clean as a Mountain Stream

Purpose

How easily can poisonous chemicals be removed from our drinking water?

Hypothesis

____

____

____

____

Materials

Large mayonnaise jar, empty
Plastic drinking glasses, 10
Water
Food coloring (red, blue, or green)
Marking pencil
Sponge

Procedures

1. Fill the mayonnaise jar with water.
2. Add 30 drops of "poisonous chemical" (food coloring) to the water.
3. Number the plastic drinking glasses from 1 to 10.
4. Fill glasses 1 and 2 half full with the "polluted water." Glass 1 is your control.
5. Add clean water to glass 2 until it is full. Compare the color of the water in glass 1 with the color of the water in glass 2.
6. Pour half the water from glass 2 into glass 3.
7. Add clean water to glass 3 until it is full. Look for any color change.
8. Repeat the procedure through glass 10.

Observations

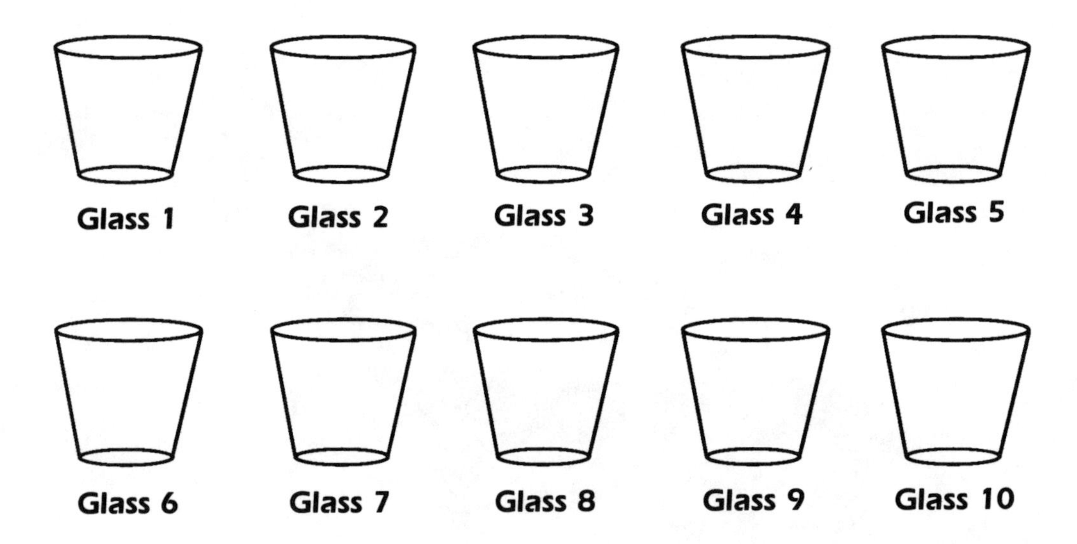

Conclusion

____

____

Name____________________ Date____________________

Layer Cake

Purpose

How are layers formed on Earth?

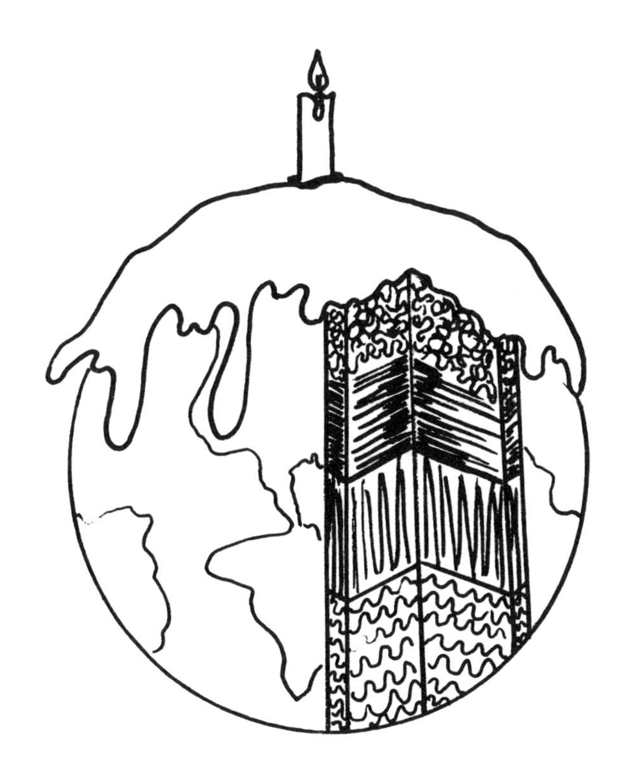

Hypothesis

____

____

____

____

Materials

Plastic peanut butter jar, empty
Small rocks
Potting soil
35mm film container, empty
Sand
Water
Clock

Procedures

1. Fill the film container with potting soil and then empty it into the peanut butter jar. Repeat this procedure five times.
2. Repeat Step 1, but use sand instead of potting soil.
3. Put a small handful of rocks into the jar. The jar should now be about half full.
4. Fill the jar the rest of the way with water.
5. Screw the lid on *tight.*
6. Shake the jar for three minutes.
7. Draw a picture of what you observe taking place in the jar.
8. Set the jar on the table and do not touch it.
9. After the particles have settled inside the jar, draw another picture of what you observe.

Observations

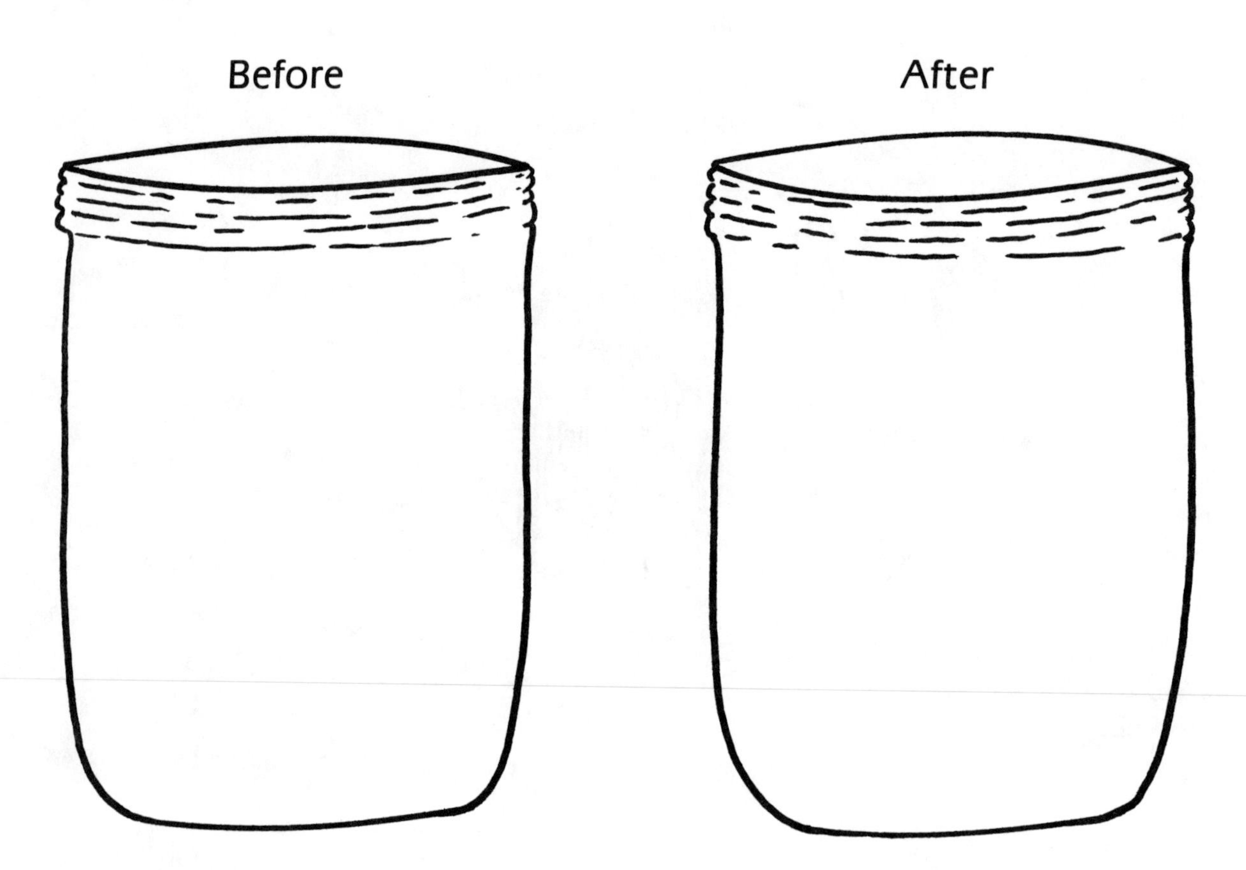

Conclusion

____

____

Student Lab Sheet 2.4

Name________________________ Date________________

Raindrops Keep Falling . . .

Purpose

Can a rainstorm be created in the classroom?

Hypothesis

____

____

____

____

Materials

Hot plate
Water
Measuring cup
Small saucepan
Aluminum cake pan
Hot pad
Ice

Procedures

1. Turn the hot plate on to its medium-high setting.
2. Put two cups of water in the saucepan. Then put the pan on the hot plate.
3. Wait for the water to get hot.
4. Cover the bottom of the aluminum cake pan with ice.
5. Hold the pan of ice about 18 inches above the saucepan.
6. Observe.

Observations

Conclusion

____

____

LAB EXPERIENCES

LIFE SCIENCE

Can You Do the Locomotion?

All it takes is one day in the classroom to know that there are many forms of movement. But how much thought have your students given to the variety of ways in which animals move? In this lab—an excellent one for refining observational skills—students will observe how several animals move and then try to imitate those movements.

Materials per Class

Chalkboard
Chalk

Preparation

Wait for a nice day outside. Review the difference between plants and animals. Discuss where students are likely to find animals on the school grounds. Review any rules you have when students are outside on school property.

Focusing Activity

Ask students to list the different ways they have seen animals move, and write their answers on the board. Use this part of the activity to build vocabulary and increase creative thinking skills. Do not label any answer as being right or wrong. The object at this point is to get students thinking about the many ways in which animals move.

Procedure for Lab

Take the class outside. Help each student find an animal to observe. If the school has a bird feeder, have some students wait near it for their observations. Encourage other students to lift rocks carefully and observe what they find underneath. Tell others that they may be surprised at what they find when they get on their hands and knees to explore the grass. Those who still cannot find an animal to observe should check the sidewalks and areas close to the building.

When each student has located an animal, he or she should observe its behavior. How does the animal move? Does it move in a straight or crooked path? Does it travel long distances or stay in the same place most of the time? Does it travel by itself or in a group? Does it stop to eat? If it does, what does it eat? What type of home does it have?

Give the students adequate time to observe, and then have them all come back together. Each student should take a turn at imitating his or her animal. Can the class guess the animal? Can the class imitate the student who observed it? Discuss the number of legs the animal has. Does the number of legs an animal has affect how it moves?

Extension Activities

1. Have the students return to the same place at various times during the day. Can they find the same type of animal they observed earlier? Does

their animal seem to prefer a specific time of day—or type of weather—to make an appearance?
2. Check out books about animals from the library and place them at a resource table. Encourage students to find out as much information as possible about their animal.
3. Have students make their own animal book, combining information from the library books with pictures of various animals that can be found on the school grounds or at home.

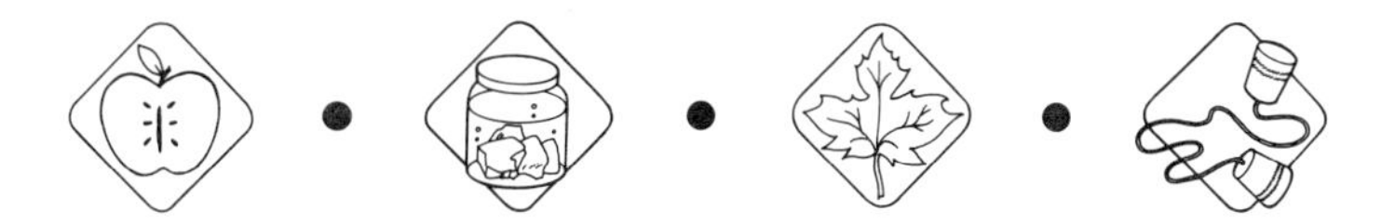

What Shape Are You In?

This lab is great for working on visual discrimination, sorting by shape, and leaf recognition. Students will explore nature to find a leaf that matches the shape you have given them.

Materials per Teacher

Construction paper
Pencil
Writing paper
Scissors

Materials per Student

Construction paper leaf

Preparation

Take a nature walk around the schoolyard, noting the shapes of leaves. You may wish to take writing paper and a pencil with you to trace the leaf shapes that you find. Use construction paper to cut shapes that match the shapes of the leaves you found.

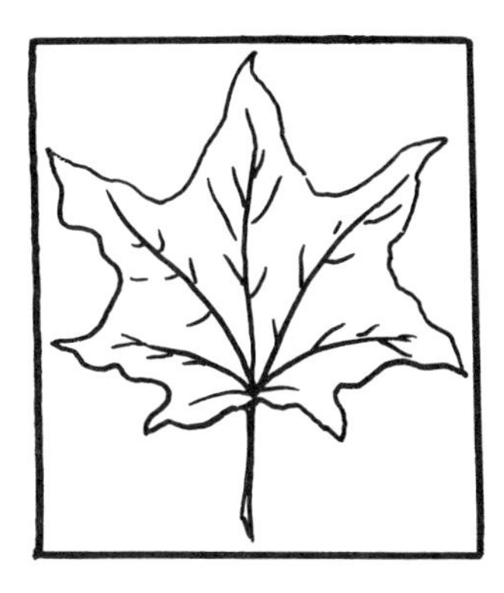
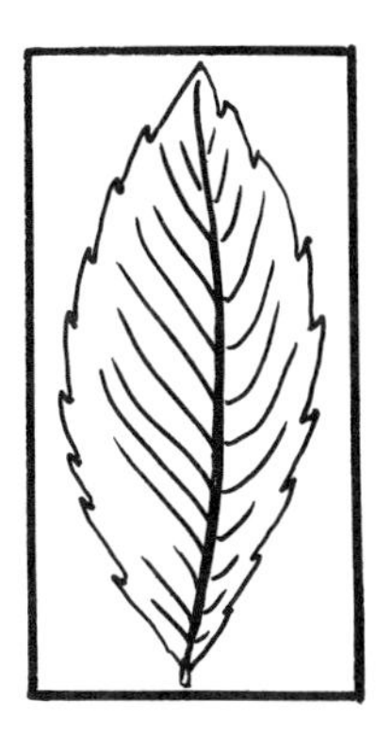

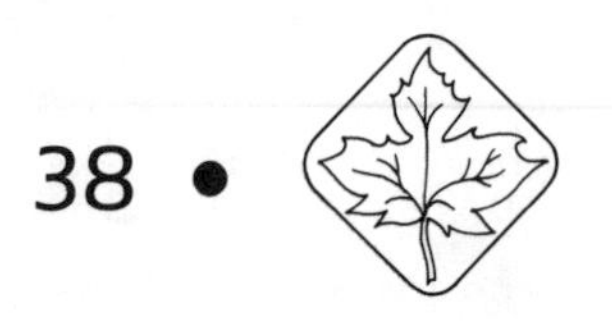

Focusing Activity

Discuss the parts of a tree—its roots, trunk, branches, and leaves. Are all trees the same shape? Are tree branches alike? What about the leaves? Are leaves all the same or are they different?

Procedure for Lab

Explain to the students that everyone will receive one construction paper leaf. They are to take that paper leaf and find the real leaf that most closely matches it. Remind students that they are *not* to remove any leaves from trees! If they find leaves on the ground, however, they may pick them up and examine them.

Tell the students that as soon as they find the leaf that matches their construction paper shape, they should stand still and put their hands on their heads. You then circulate among the students and have them prove that they indeed found a matching shape. You may wish to have students compare their leaves.

Extension Activities

1. Go on a nature walk through a wooded area, collecting fallen leaves. Place the leaves at a learning center, along with a tree book. Can students identify the leaves? Have them group the leaves by approximate size, count the number of leaves in each group, and then graph the results. They can also rank the leaves in order from smallest to largest and classify the leaves according to color or shade.
2. Dry, press, and preserve leaves. Use the preserved leaves to make a leaf book that students can look through.

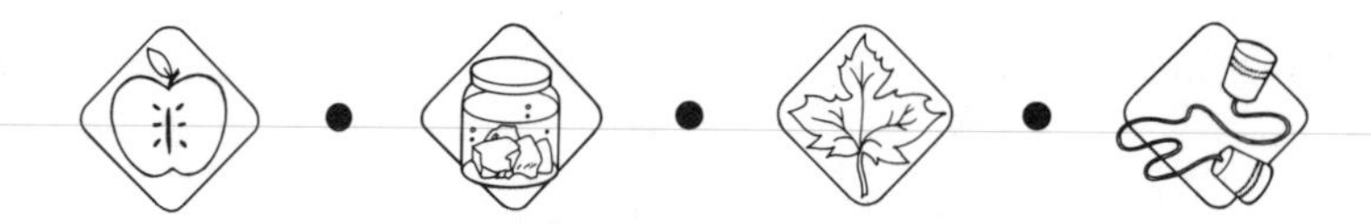

The Ants Go Marching One by One!

Children tend to be fascinated by anything that is very small. You can capitalize on that fascination by taking students on a special hike. In this lab, students pretend to be the size of an ant. As they examine the world from that perspective, students sharpen their observational skills. This lab can also provide opportunities for students to improve their graphing abilities and animal-classification skills.

Materials per Student

String, 3 feet
Magnifying glass
Student Lab Sheet 3.1
Pencil

Preparation

Cut string into three-foot sections. Duplicate Student Lab Sheet 3.1. Wait for a nice day outside to conduct this activity.

Focusing Activity

Explain to the students that they will be studying animals in today's lab. Since the best way to study animals is to see them in nature, the class is going to take a hike. But this hike is different from any other they've ever taken; this one is a micro-hike.

Ask students if they know what the prefix "micro-" means. If not, introduce a few words that begin with the prefix (e.g., microscope), and have one student look up micro in the dictionary and read the definition to the class. Conclude by saying that not only the hike will be tiny; so will the hikers!

Procedure for Lab

Hand out the pieces of string. Explain to the students that the string will mark the trail they will hike along. They can lay the string on the ground outside anywhere they like, but they should try to choose an interesting location.

Then tell students that when hiking along their trail they must pretend that they have been reduced to the size of an ant. To help students conceptualize how long their three-foot path is compared to their small size, tell them the distance would be the same as if they were walking from one end of a football field to the other . . . twice! Remind the students not to travel too quickly or to disturb anything along their trail. As they investigate, they should record their observations on their lab sheet chart. For example, if they find an animal during their hike, they should count and record the number of legs it has without disturbing it.

Begin the lab by waving your hands over your head and saying a magic word. Tell the students, now all reduced to the size of ants, to go outside quietly and lay down their string trail.

Circulate among the students as they are "hiking." Ask thought-provoking questions such as: Do they see a possibility for lunch? Might some other creature on the trail see *them* as lunch? How does it feel to be so small? What obstacles are blocking their path? Do they see any animal homes? Where might they stop along the trail for shelter? Remind them to record any animals they see by using tally marks on their lab sheet chart.

Gather the exhausted hikers and return quietly to the classroom. Discuss with the students what they saw along their trails, and then—using the information from their lab sheet—have them make a bar graph.

Extension Activities

1. Repeat the micro-hike, but this time have students classify the animals they see along their trails.
2. Go on a nature walk. Remind students to be very quiet so as not to disturb any wildlife they may see.
3. Create a large graph that combines everyone's micro-hike information.
4. Set up an ant farm at a learning center. Remember to have resource books and a magnifying glass close at hand.
5. Teach the song, "The Ants Go Marching One by One."

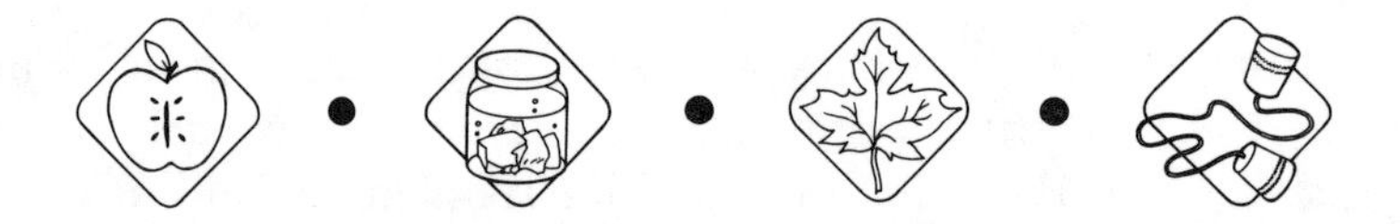

Looking for the Straight and Narrow

In this lab, students will determine whether there are more monocots or dicots in the schoolyard. One method for determining if a plant is a monocotyledon or a dicotyledon is to examine the seed. But what can you do if no seeds are present? Examine the leaves! This lab gives students experience in studying the veins of various leaves, and it reinforces the correct usage of two prefixes: mono- and di-.

Materials per Lab Team

Whole peanut
Popcorn kernel
Lima bean

Materials per Student

Student Lab Sheet 3.2
Pencil
Clipboard or a piece of cardboard to hold lab sheet

Preparation

Walk around your schoolyard and look at the leaves of various plants. Locate examples of monocotyledons and dicotyledons. Duplicate Student Lab Sheet 3.2.

Focusing Activity

One of the easiest ways to distinguish a monocot from a dicot is by examining the seeds. Monocotyledons have one food packet in their seeds to feed the plant embryo. The seeds do not split apart. A good example would be a corn kernel. Dicotyledons, in contrast, have two food packets in their seeds to feed the plant embryo. Dicotyledon seeds split easily into two nearly identical pieces. A good example is a peanut.

Give each lab team a lima bean that has been soaked in water for several hours and a popcorn kernel. Explain that scientists classify members of the plant kingdom just as they classify members of the animal kingdom. One way to classify green plants is to divide them into monocotyledons and dicotyledons. The root word "cotyledon" means food packet.

Ask students what the prefixes "mono-" and "di-" mean. If necessary have a student read the definitions from the dictionary. Ask how many food packets a dicot has. Explain that the seed from a dicot will split into two almost identical pieces; when the embryo starts to grow from the seed, it will have two leaves on the seedling.

Have the lab teams examine the two seeds you gave them. Which is the dicot? The lima bean. How do they know? The lima bean can be split easily into two pieces. Have the students look carefully at the plant embryo inside the seed; they will see that it does indeed have two leaves on it.

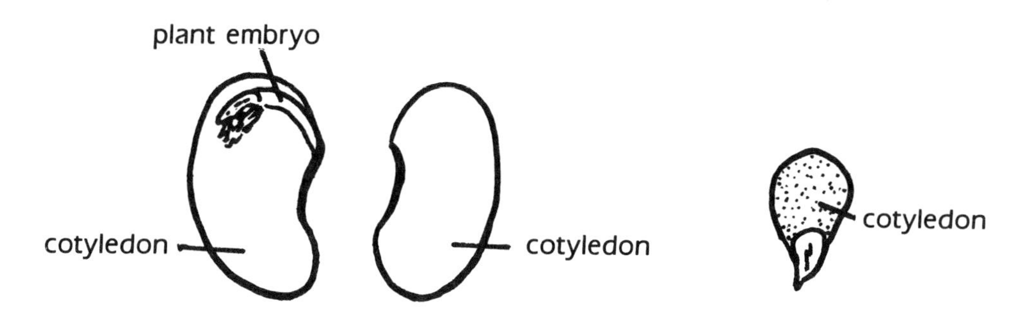

Opened Lima Bean
Dicotyledon

Popcorn
Monocotyledon

Ask how many food packets a monocot has. Explain that the seed from a monocot will not split into pieces—it stays intact; when the embryo starts to grow from the seed, it will have one leaf on the seedling. Have the lab teams examine the popcorn kernel. How do they know that it is the monocot? It cannot be split.

Procedure for Lab

Ask students how they can tell a monocot from a dicot. They should respond that the seeds are different. A dicot has two packets while a monocot has one food packet.

Then ask how they could tell a monocot from a dicot if they did not have any seeds from the plants. Instruct students that monocots and dicots also differ in their leaves. The veins in a monocot leaf run parallel to each other. The veins in a dicot leaf are spread out in many directions, like a web. Again, refer students back to the prefixes. In a *mono*cot leaf, the veins run in just one direction.

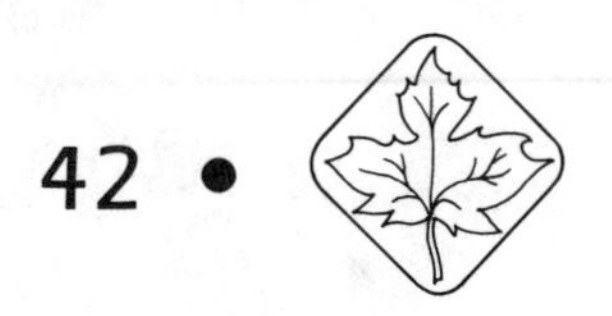

Dicotyledon
"Dicot"

Monocotyledon
"Monocot"

Explain that lab teams will go into the schoolyard to figure out whether there are more monocots or dicots on the school grounds. If they find a monocot, they should put a tally mark on the chart under the "Monocot" heading. If they find a dicot, they should put a tally mark on the chart under the "Dicot" heading. Remind students to make sure they count each plant only once and *not to pick the leaves off any plants!*

For best results, limit the outdoor counting time to ten minutes so that students stay on task. When they return to the classroom, have them share their findings. Are there more dicots or more monocots? Do the teams basically agree in their counts? Discuss any significant differences. Did each group examine the same area? Does everyone know how to identify dicots and monocots?

Extension Activities

1. Make individual leaf rubbings. Use one piece of paper for monocots and another for dicots.
2. Have the students repeat the lab exercise at home and then compare their findings with the count of dicots and monocots at school.
3. Create a class graph of the total number of monocots and dicots on the school grounds.

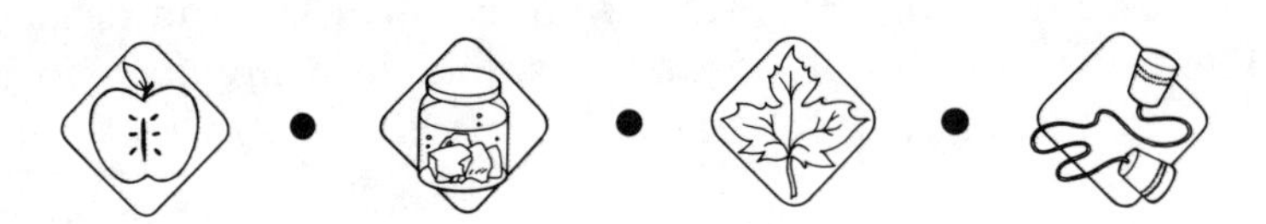

Hide and Seek

How does the game of hide and seek work in the animal kingdom? In this lab, students will learn how animals use their surroundings to protect themselves. They will learn the meaning and importance of camouflage, and they will have an opportunity to practice their graphing skills.

Materials per Class

Toothpicks, box
Food coloring, red and green
Bowls, 2
Paper towels
Watch
Chalkboard
Chalk

Materials per Student

Plastic bag
Crayons, red and green
Student Lab Sheet 3.3
Pencil

Preparation

Break each toothpick into thirds. Divide the toothpick segments into two equal piles. Put three drops of red food coloring in one bowl and three drops of green food coloring into the other. Add a small amount of water to each bowl. Next, add one pile of toothpicks to each bowl and mix them around. If possible, allow the toothpicks to soak for one hour. Drain off the water and allow the toothpicks to dry on paper toweling. When no students are around, scatter all the toothpick segments over a section of grass on the schoolyard. Be sure to note the boundaries of the area you use.

Focusing Activity

If possible, come to class wearing a camouflage outfit. If that isn't possible, bring in a toy soldier wearing one.

Ask students what animals do to protect themselves from their enemies. Encourage students to list as many ways as possible. Correct responses include using quills, sharp claws, loud growls, odors and other chemicals secreted from the body.

When a student responds that animals also use protective coloring, ask the class why you (or the toy) are wearing a camouflage outfit. Wait for all responses.

Procedure for Lab

Explain that today's lab concerns how camouflage works. Tell the class that you have discovered an area of the schoolyard that is filled with "critters."

Some of the critters are red while others are green. Hold up a red toothpick segment and a green one, and explain that these are what the critters look like. Now tell the students to pretend that they must have the red and green critters in order to survive. Inform the class that there are as many red critters as green critters on the schoolyard. Then ask the class which kind of critter they think they can find more of. Why?

Give each student a plastic bag. Explain that they are to put every critter they find in the bag and carry all of them back to the classroom. Tell them not to pick up any critters until you say "Go" and to end their critter search immediately when you say "Stop."

Take the class outside and identify the search boundaries. Note that instead of listening to what you have to say, students will be eagerly scanning the ground looking for critters. Ask if there are any questions. Answer the questions, say "Go," and then stand back. Allow the search to go on for five minutes, and then say "Stop." Although some students will beg to find "just one more," be firm. Lead the critter-toting students back inside.

Give each student a piece of graph paper. Explain that they should construct a bar graph that illustrates the results of their search. Point out that the students will have to label the x-axis and y-axis. Along the x-axis, students must label the columns of the graph correctly. Along the y-axis, students must place appropriate numbers— ones, twos, fives, tens, etc.— depending on how many critters they've captured.

When the students have finished their graphs, discuss which critter was the most difficult to find, the red or the green. Why? Students should respond that the green critter had the better camouflage and was better able to hide among the blades of grass.

Have the students discuss their results and create a composite graph of all their findings on the chalkboard.

Extension Activities

1. Place library books that deal with camouflage at a learning center for students to use.
2. Repeat the experiment, but this time scatter the toothpicks over an area of asphalt rather than grass. Are the results the same?
3. Divide the class into two teams. Have each team make a critter out of clay and then create camouflage for it. Allow each team to hide its critter and search for the other. The first team to find the other team's critter is the winner.

Student Lab Sheets

LIFE SCIENCE

Name____________________ Date________________

The Ants Go Marching One by One!

ANIMAL TALLY CHART

DESCRIPTION	NUMBER FOUND
Animals with no legs	
Animals with 2 legs	
Animals with 4 legs	
Animals with 6 legs	
Animals with 8 legs	
Animals with more than 8 legs	

ANIMAL GRAPH

NUMBER OF ANIMALS

13						
12						
11						
10						
9						
8						
7						
6						
5						
4						
3						
2						
1						
	ANIMALS WITH NO LEGS	ANIMALS WITH 2 LEGS	ANIMALS WITH 4 LEGS	ANIMALS WITH 6 LEGS	ANIMALS WITH 8 LEGS	ANIMALS WITH MORE THAN 8 LEGS

DESCRIPTION OF ANIMALS

Name________________________ Date________________

Looking for the Straight and Narrow

Look closely at the leaves of plants on the schoolyard. Figure out whether there are more monocots or more dicots. For every monocot you find, make a tally mark (see the example below) under the heading "Monocots." For every dicot you find, make a tally mark under the heading "Dicots."

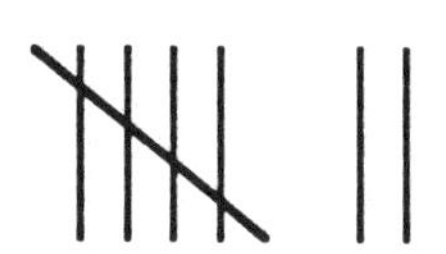

Dicotyledon
"Dicot"

Monocotyledon
"Monocot"

MONOCOTS	DICOTS

Name______________________ Date______________

Hide and Seek

CRITTER GRAPH

NUMBER FOUND

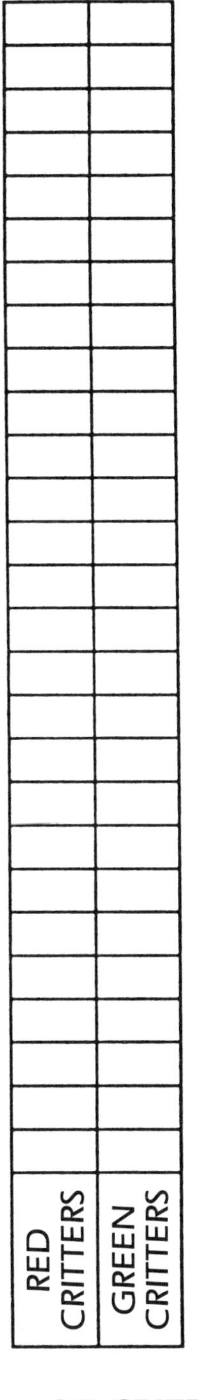

TYPE OF CRITTER

LAB EXPERIENCES

PHYSICAL SCIENCE

 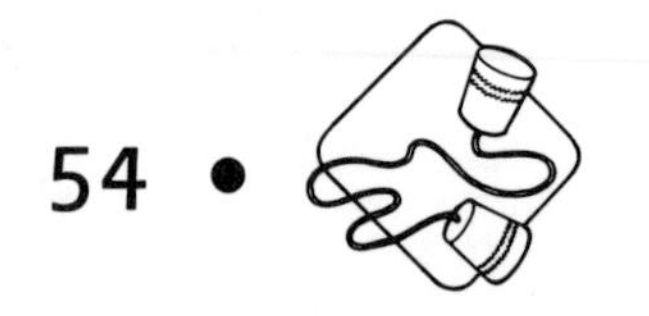

An Enlightening Experience

How many times a day does a student flip on a light switch without ever thinking about the concept involved? This is an introductory lab involving series circuits. Using a battery, bulb, and wires, students will discover how to make the bulb light up. They will also find that when the circuit is broken, electricity will not flow.

Materials per Lab Team

D-size battery
Small bulb (e.g., flashlight bulb)
Insulated wire, 2 pieces (or alligator wires, 2)

Materials per Student

Student Lab Sheet 4.1
Pencil

Preparation

Test all the supplies prior to the lab. Nothing frustrates students more than to do everything correctly and then discover at the end of the lab that the bulb does not light because one of the parts is malfunctioning.

If you are using wire, strip the insulation back from each end about one centimeter. Duplicate Student Lab Sheet 4.1.

Focusing Activity

This lab is discovery oriented. Students will progress at their own speed through the lab sheet, recording their observations as they go.

It is important to discuss the dangers of electricity with the students. Many have been told from an early age that electricity is dangerous (it is), and they bring their fear with them into the lab. Emphasize, however, that the batteries involved in this lab cannot hurt them.

Go over the drawings on the Student Lab Sheet so that students understand the diagrams. Have the students compare the illustrations with the actual objects. Point out the position of the battery; students often run into difficulty in this lab because they have the battery turned around.

Procedure for Lab

Once students understand the diagrams, allow them to progress through the lab sheet at their own pace. Walk among the lab teams, reminding students to write their hypothesis before testing the circuit. Ask, "Why do you think that the light bulb lights/does not light? What do the circuits that work have in common? Does the position of the light bulb affect its ability to light? Does the position of the battery affect the bulb's ability to light?"

When all teams have completed the lab sheet, have them compare results. The bulb should light in problems 1, 2, 3, 7, and 10. If there are any discrepancies, have the students retest their circuit. Explain that when the light bulb lighted, the battery and bulb were wired in series. Ask students

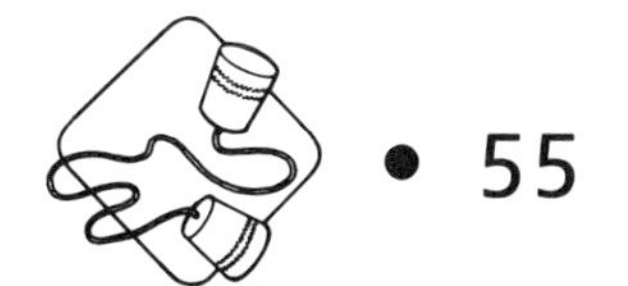

what they think of when they hear the word "series." Expect at least one student to respond "the World Series" in baseball. Point out that in the World Series one game is played after another.

Have the students join hands and form a circle. Explain that a series circuit is much like this circle. Have students pretend that electricity is passing between their hands. What would happen if two students dropped hands? The circuit would not be complete, and electricity could not continue to travel. That's what happens in a series circuit if one of the component parts fails to work or becomes disconnected, breaking the circuit.

Extension Activities

1. Contact your local electric utility company. Many supply educational programs and materials for use in schools.
2. Conduct an experiment to find out what effect another battery in series has on a light bulb (the bulb becomes brighter) and whether the direction in which the batteries are placed makes a difference (the batteries must be wired in the same direction for the circuit to work).
3. Conduct an experiment to find out what effect another light bulb has in a series circuit. The bulb becomes dimmer.
4. Help lab teams build a lamp. Parts are available from hardware stores, or you may even be able to obtain the parts you need from an electrical contractor's scrap pieces.
5. Have students design and build electric quiz boards using cardboard, insulated wire, a battery, light bulb, and metal rings.
6. Encourage students to write an account of how they would get through a day without electricity.
7. Have students prepare a report on Thomas Edison and his many electrical inventions.

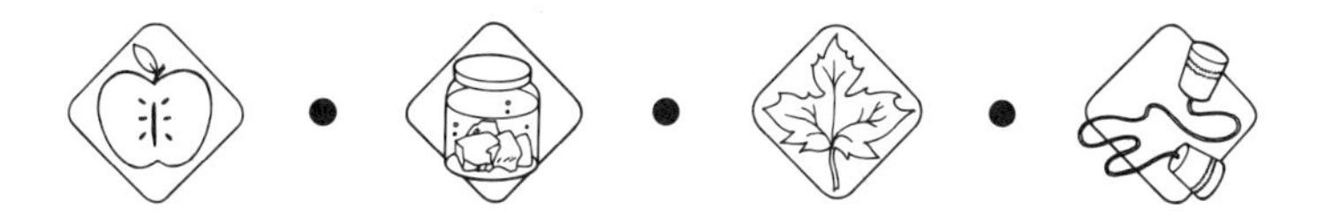

Under Pressure

Air exerts a pressure of 14.7 pounds per square inch at sea level. Why don't we feel this pressure? In this lab, students will observe a dramatic display of air pressure and draw conclusions about equal and unequal forces.

Materials per Class

- Hot plate
- Metal ditto fluid or gas can THOROUGHLY CLEANED!
- Water
- 1-cup measuring cup
- Tongs

Materials per Lab Team

Plastic tumbler
Water
Index card larger than the opening of the tumbler
Sink or bucket

Materials per Student

Student Lab Sheet 4.2
Pencil

Preparation

Rinse the ditto fluid or gas can thoroughly. Any flammable liquid remaining in the can during this lab could start a fire!

Focusing Activity

Use a bit of creative dramatics to introduce the concept of air pressure. Walk around the room bent over and complaining that you can feel a terrible push against your body. Periodically look over your shoulder as if trying to catch a glimpse of what might be creating this pressure. Ask students what they think is pushing against your body. Allow time for responses and discussion.

Procedure for Lab

Explain that today's lab consists of two parts. The first part is a class demonstration. The second part is a hands-on experiment. Make clear to the students that they must pay careful attention to the procedures for the demonstration because they are not written on the lab sheet. Tell the students that they will be expected to fill in the steps themselves on the lab sheet, using the correct format. Take a few minutes to review how procedures are written in numeric sequence.

To encourage class involvement during the demonstration, have individual students perform each step prior to the heating of the can.

Turn the hot plate to high. Uncap the thoroughly rinsed metal can. Discuss why it was necessary to rinse the can prior to the lab. Measure and pour one cup of water into the can. LEAVE THE CAN UNCAPPED! IF THE CAN IS HEATED WITH ITS CAP ON, IT COULD EXPLODE.

Place the can on the hot plate. While it is heating, students should write all the procedures just completed and draw a "Before" picture depicting the can on the hot plate. Allow the water to boil. As the steam is escaping, ask students to explain what is happening. They should know what is leaving the can (water vapor and warm air) and why it is leaving (the heated air and water molecules are spreading farther apart). In order to encourage students to make detailed observations, allow this part of the lab to continue for five to ten minutes before going to the next step.

Then use the tongs to remove the can from the hot plate, and *quickly* cap the can. Ask students to observe the results (the can begins to collapse very shortly after it is capped) and explain why it was necessary to cap the can quickly. Then have them complete the written procedures on their lab sheets and draw an "After" illustration.

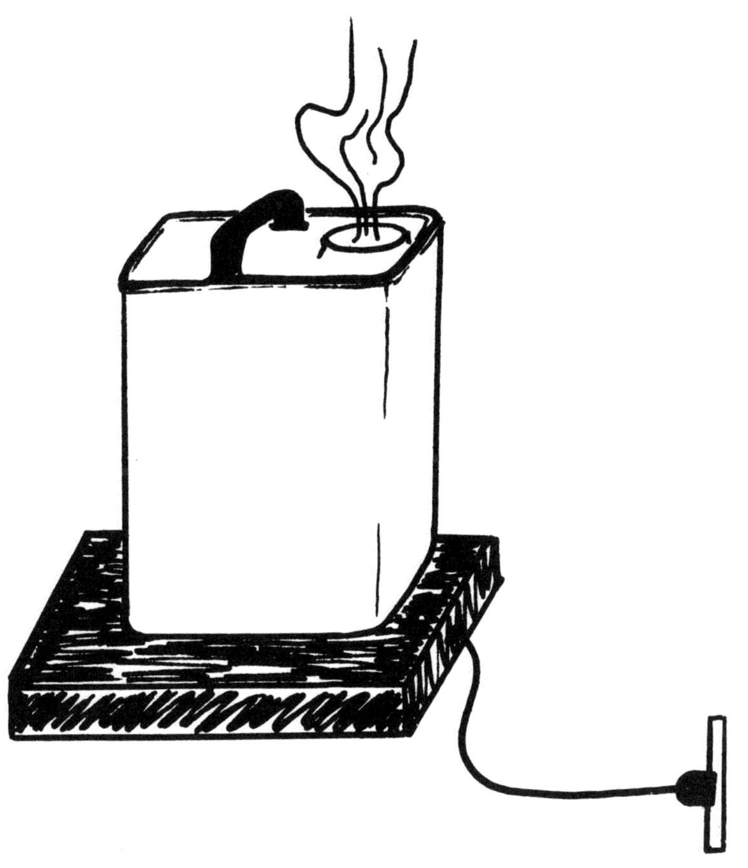

Discuss the reason for the can's collapse. Where was the air pressure higher and where was it lower? Students should respond that it was higher outside the can and lower inside the can. They should then write their conclusions explaining *why* the can was crushed.

Ask the students what they might infer would happen if the air pressure were reversed—i.e., greater inside the can than outside. They should respond that the can would explode. Explain that it is for precisely that reason one should never heat a closed container.

Next, have the lab teams do an experiment using the tumbler, water, and index card. Tell them to be sure to do the experiment over a pail or sink. Have paper towels ready!

Students should first fill the tumbler to the top with water and then set an index card over the top so that it covers the entire opening. With one hand on the bottom of the tumbler and the other on top of the card, students then invert their tumblers and slowly withdraw the hand touching the card. The card should stay on the tumbler, holding the water inside.

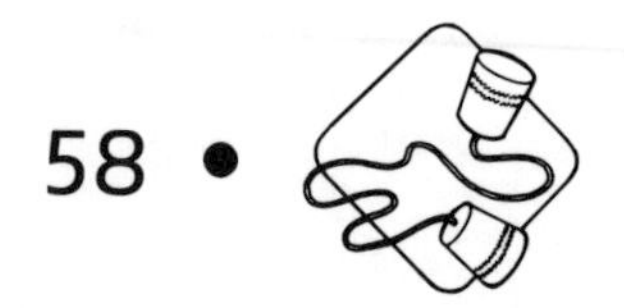

Have the students discuss the reason for this phenomenon. They should understand that the air pressure pushing up on the index card is greater than the pressure pushing down on the water.

Extension Activities

1. For an even more dramatic demonstration, repeat the tumbler/water/card experiment using a stiff piece of wire screen material instead of the index card. Make sure the spacing between the wires in the screen is very small.
2. Hold a contest in which students compete to discover the amount of air pressure exerted on the body. The winner must be able to reference his or her correct answer.
3. Have students brainstorm to create a list of the positive uses of air pressure (e.g., inflate tires).
4. Make a bulletin board featuring pictures (brought in by students) that show the effects of air pressure.
5. Invite a local meteorologist to discuss barometric pressure with the class.
6. Have students keep track of the local barometric pressure (via TV or newspaper reports), and use that information to create a giant chart in the classroom. Discuss how the barometric pressure affects the weather.

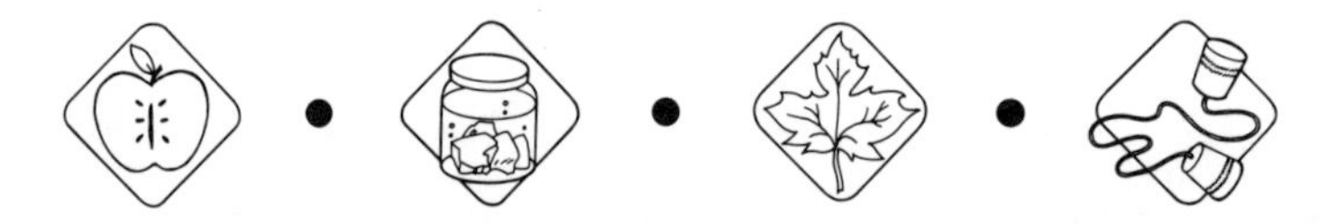

May the Force Be With You!

During this lab, students will investigate how changing the position of a fulcrum changes the amount of force required to do work. They'll discover that moving the fulcrum closer to the load reduces the amount of force required.

Materials per Lab Team

Pencils, 3
Wooden metric ruler
Tape, either cellophane or masking
Small ball of clay
Washers or pennies, 10

Materials per Student

Student Lab Sheet 4.3
Pencil

Preparation

Students should be familiar with the terms "lever," "fulcrum," "load," "force," and "work." Duplicate Student Lab Sheet 4.3.

Focusing Activity

Walk in front of the class hunched over with your hand on the small of your back. Groan as if you were in pain. Once you have the students' attention, explain that you hurt your back while trying to lift a heavy load. Ask students if they know of any simple machines that make it easier to lift heavy loads. Allow students time to respond. When someone mentions a lever, review the terms "fulcrum," "load," "force," and "work."

Ask the students whether the position of the fulcrum affects the amount of force needed to do work. Have them hypothesize an answer to your question.

Procedure for Lab

This lab is good for students gaining practice in following procedures on their own. Because all the students may not assemble materials the same way, this lab provides an excellent opportunity to discuss why different lab teams get different results. Students should understand that not setting up the experiment in the same manner will produce different results.

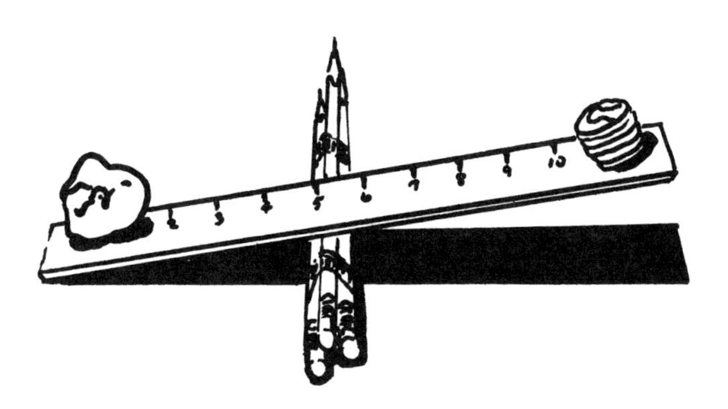

As students measure the amount of force required, they should record their findings on their lab sheets. Remind them that when they record measurements they must do so in the unit of measure specified.

Circulate among the lab teams during the experiment, reinforcing the meanings of the terms "lever," "fulcrum," "load," "force," and "work." Have the students demonstrate the definitions by pointing to relevant points on their seesaw.

Students may experience difficulty with Step 6 of the procedure. Usually the trouble results from placing the fulcrum at the 8 centimeter mark on the ruler instead of 8 centimeters from the load. If a student makes this mistake, try to turn it into a learning experience. Ask everyone to reread the procedures and try to figure out what could have gone wrong.

At the end of the experiment, have lab teams discuss their results. Emphasize the importance of using the word "because" in the conclusion to avoid merely restating observations. A conclusion should tell why something happened, not just what happened.

Students should conclude that the force needed to lift the heavy load

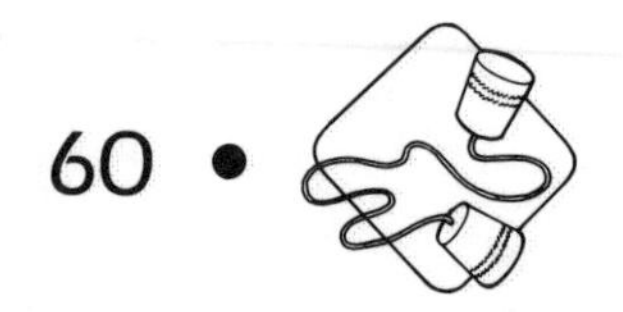

does not change, regardless of the position of the fulcrum. When the fulcrum is closer to the load, the distance one must push the lever to lift the load is greater. The fulcrum in this instance spreads the force over a greater distance. When the fulcrum is farther from the load, the distance one must push the lever to lift the load is smaller. In this instance, the fulcrum concentrates the force.

To illustrate this principle, take two slices of bread of vastly different sizes. The bread represents the distance the lever must travel. Measure out a tablespoon of peanut butter for each slice of bread. The peanut butter represents the force. To cover the large slice of bread, the peanut butter must be spread very thin (just like the force spread over the greater distance). To cover the small slice of bread, the peanut butter must be spread very thick (just like the force spread over the smaller distance).

Extension Activities

1. Assign students to search the playground at school for evidence of simple machines (especially the lever) in use.
2. If the school playground has a seesaw, carry out an experiment by moving the fulcrum. Have the students write a report about the experiment in which they state how moving the fulcrum affected the seesaw.
3. Encourage students to find other examples of levers and to share their findings with the class. Among the common levers students are likely to mention are the bottle opener, car jack, scissors, and tongs.

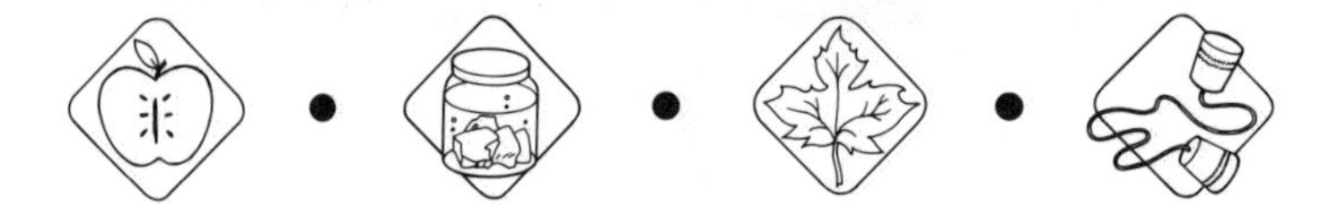

I Can't Hear You!

How many times have your students told you, "I didn't hear you say that!"? In this lab, they will discover that sound travels better–and, therefore, is louder–when it travels through a solid substance rather than through a gas.

Materials per Lab Team

Bathroom-size paper cups, 2
Paper clips, 2
String, one 72-inch piece and two 24-inch pieces
Coat hanger

Materials per Student

Student Lab Sheet 4.4
Pencil

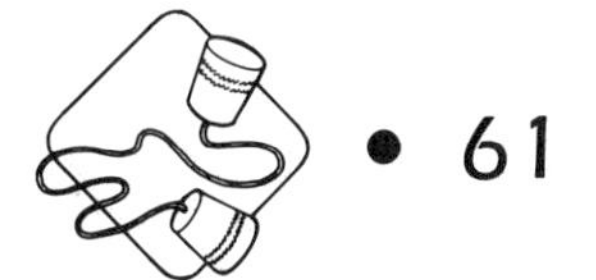

Preparation

Students should know the difference between a solid and a gas prior to this lab. They should also know that sound is caused by vibrations moving through the air.

Since the actual set-up for this lab is relatively simple, it's a good one for students to try doing themselves. Duplicate Student Lab Sheet 4.4.

Focusing Activity

To get students' attention, announce that it is starting to snow (especially effective on a warm, sunny day!). After the students have stopped looking out the windows, ask them to repeat what you just said. How do they know that's what you said? How did your voice get to their ears? Try to get them to realize that the sound produced by your voice traveled through the air to their ears.

Procedure for Lab

Ask students if they think sound travels better through a gas (such as air) or through a solid (such as a table or string). Direct students to their lab sheets, and point out the purpose of today's lab. Tell students that they should write down their hypothesis, and remind them that a hypothesis is an educated guess that answers the purpose. Because it is a guess, the hypothesis may turn out to be wrong; but it is fine for a hypothesis to be proven wrong! Some students may want to go back and correct an hypothesis after the experiment. Convince such students that many scientists make an incorrect hypothesis and go on to make very important discoveries.

When students finish writing their hypothesis, draw their attention to the list of materials each lab team will need. Tell them to choose one person from the team to gather the team's materials. Gathering materials from the lab sheet list provides good practice in reading and assimilating information.

When each lab team has its materials, instruct the students to tie the two shorter strings to the hanger, one string at each of the bottom corners. Next, have them construct a string telephone by connecting the two paper cups with the long string, secured by the two paper clips. All that they need do is punch a small hole in the bottom of each cup (a pencil point does the job well), thread one end of the string through each cup bottom, and tie the paper clips onto the ends of the string to prevent the string from being pulled back through the holes.

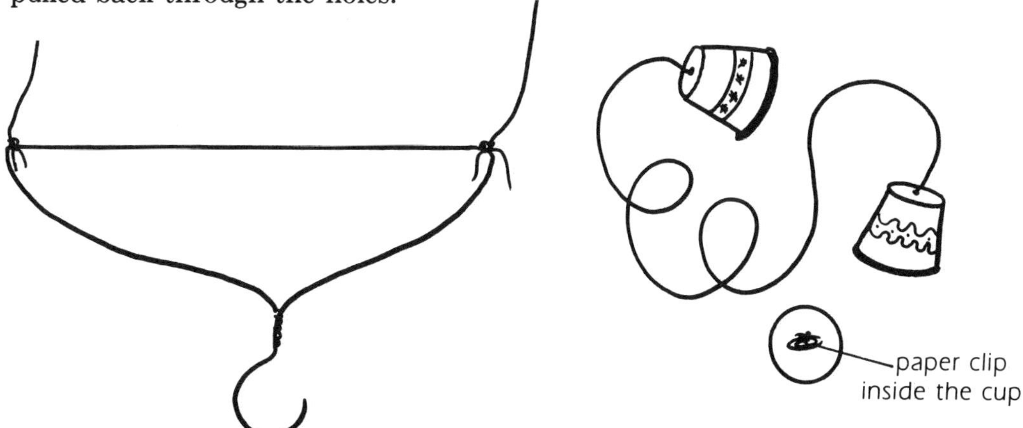

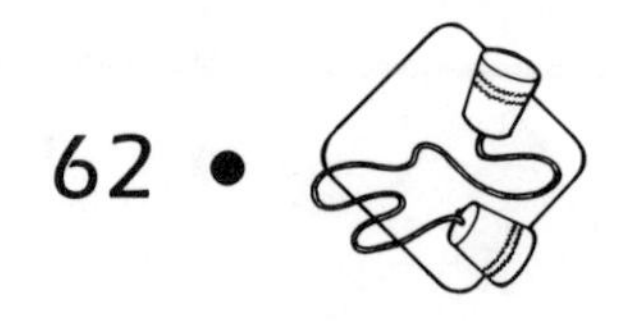

Instruct the students to follow the procedures listed on the lab sheet and then to record their observations. This provides an excellent opportunity to check reading comprehension. When all the teams have completed the experiment, discuss the results. Did the sound travel better through the air or through the solid? Why? Students should understand that the sound traveled better through the solid because the molecules of the solid are closer together, making it much easier to pass the vibration. The molecules in a gas are spaced farther apart, making it more difficult to pass the vibration.

Extension Activities

1. Have students design and carry out a similar lab in which they determine whether sound travels better in a liquid.
2. Encourage students to create pictures of sound waves traveling through different substances.
3. Initiate a unit about the human ear that includes some study of hearing problems. Have students investigate sign language and its usefulness to people who have severe hearing loss.
4. Create a matching game by cutting out pictures of animals and then clipping off the ears. See if students can match the animal with its ear.

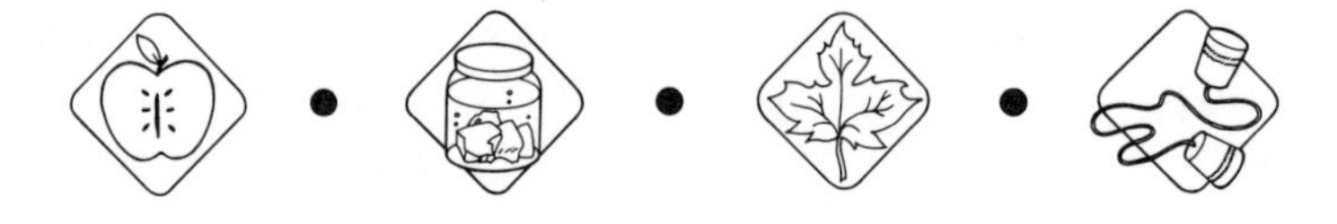

And the Winner Is . . .

Students love contests, and this lab will tap that love as well as their creativity while teaching the scientific concepts associated with insulation. It will also reinforce their knowledge of states of matter as students try to prevent an ice cube from melting.

Materials per Lab Team

Ice cube (same size for each team)
Insulation materials (e.g., sand, Styrofoam, aluminum foil, paper, cardboard, fabric, plastic)

Preparation

Give students advance notice of the experiment so that they can bring some of the insulation materials from home. It's important, however, that no team have too much of any one kind of insulating material so that students are forced to use their creativity.

Focusing Activity

Prior to starting this lab, have students submit a paragraph in which they describe how they plan to prevent their ice cube from melting and what supplies they will need. If a request cannot be filled, inform the student immediately so that he or she can revise the list of needed supplies. This is also a good time to make the point that scientists generally plan experiments carefully. The student paragraphs provide practice not only in scientific planning but also in refining writing skills.

Procedure for Lab

Place four students in charge of distributing the insulation materials, but make sure that all students receive the materials they requested in their paragraphs.

Try to capture the spirit of a contest as you circulate among the students while they construct their "ice keepers." Ask why they selected certain materials as insulation and what problems they think they might encounter. Remind the students that they must build a keeper that will allow the ice cube to be checked several times during the day.

When everyone has finished building an ice keeper, pass out the ice cubes and let the contest begin! Check the condition of the cubes several times during the day and determine which one is the winner.

After the contest, discuss what the students have learned. What could they have done differently to make their cubes last longer? What materials seemed to work well at insulating? Which didn't work well at all? Students should understand that the purpose of insulation is to work as a barrier that prevents the transfer of heat. Relate this purpose to the use of insulating materials in homes.

Extension Activities

1. Discuss how temperature changes matter. Students should understand that colder temperatures cause molecules to come close together while warmer temperatures cause molecules to move farther apart. Molecules are closest in a solid, farthest apart in a gas.
2. Plan a meal that shows how matter can change. Make pancakes (liquid to solid) with melted butter (solid to liquid) and orange juice from frozen concentrate (solid to liquid).
3. Hold an ice-cube melting contest in which the first ice cube to turn completely from solid to liquid is the winner.
4. Read *The Berenstain Bears' Science Fair* by Stan and Jan Berenstain. It covers solids, liquids, and gases in the unique Berenstain fashion.

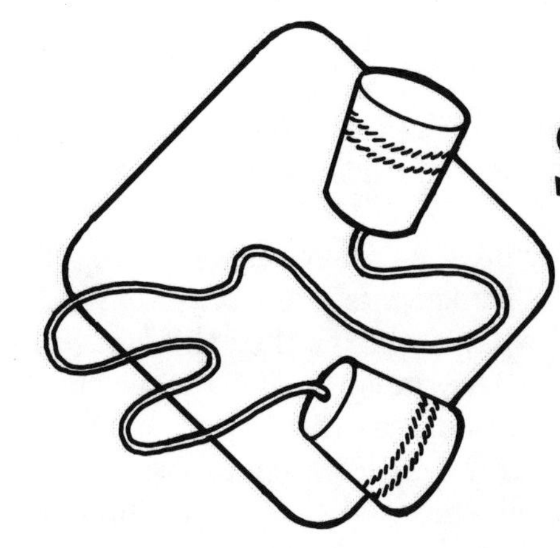

Student Lab Sheets

PHYSICAL SCIENCE

Student Lab Sheet 4.1

Name________________________ Date________________

An Enlightening Experience

Write your hypothesis before testing your circuit. If you think the light bulb will *light*, put the letter "L" on the line under the hypothesis heading. If you think it will *not light*, put the letters "NL" on the line.

After you test the circuit, write what happened on the line below the result heading.

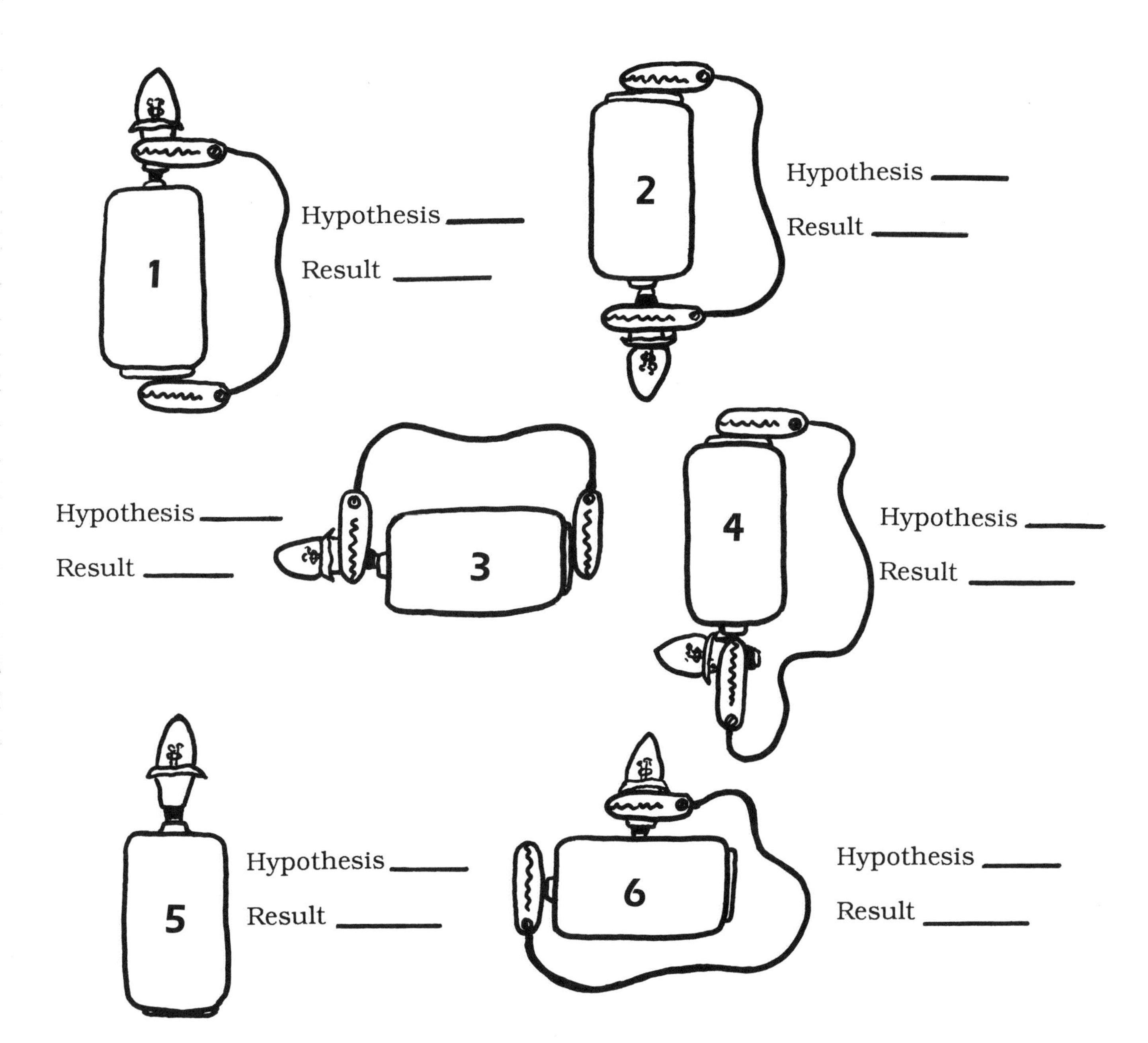

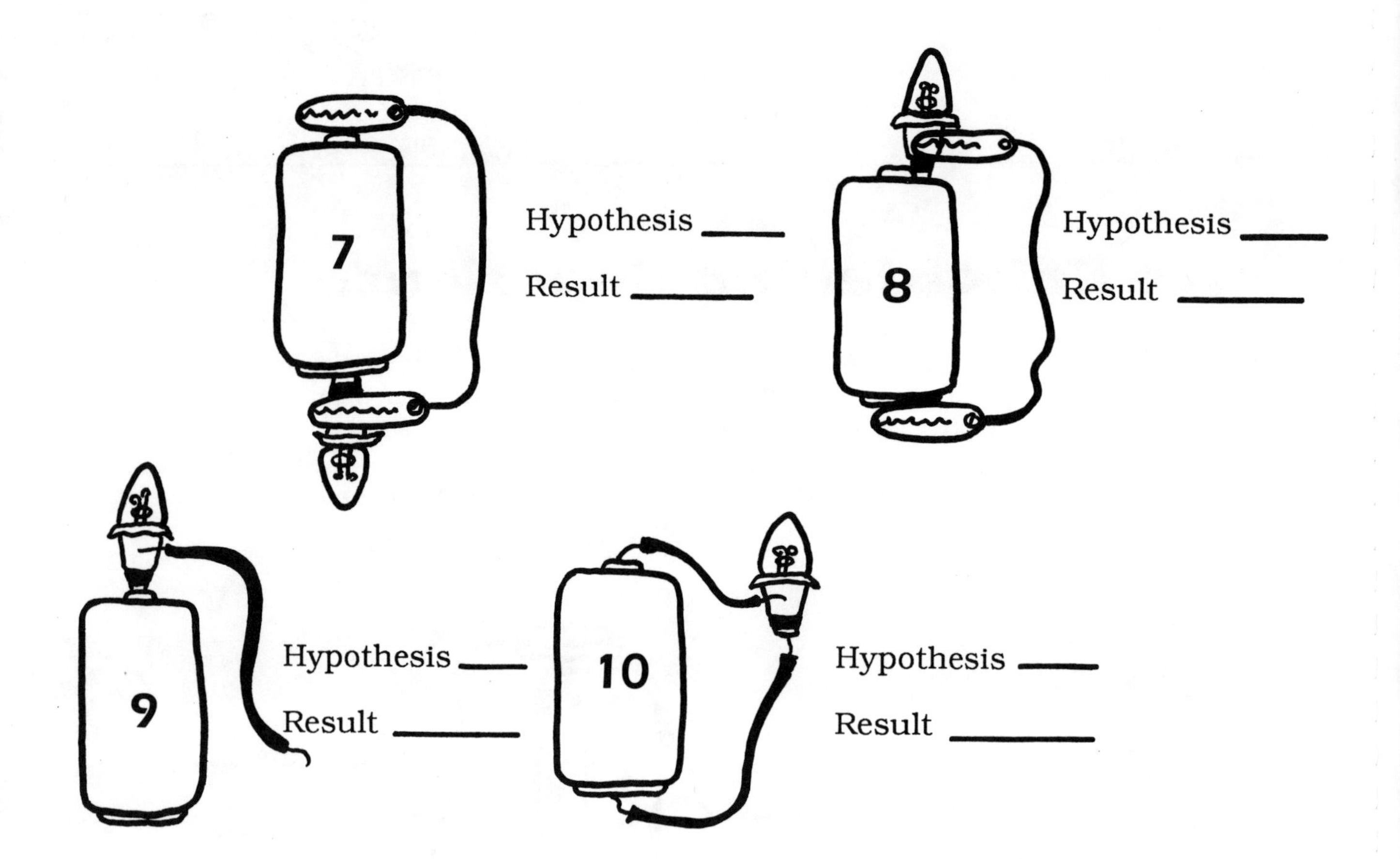
7
Hypothesis ____
Result ______
8
Hypothesis ____
Result ______
9
Hypothesis ____
Result ______
10
Hypothesis ____
Result ______

• Student Lab Sheet 4.2

Name________________________________ Date____________________

• Under Pressure

Purpose

What effect does air pressure have on a metal can?

Hypothesis

____

____

____

____

Materials (list everything needed for the experiment)

____________________ ____________________

____________________ ____________________

____________________ ____________________

____________________ ____________________

Procedures (list all the steps in the correct order)

____

____

____

____

____

____

____

____

____

____

____

____

Observations

BEFORE	AFTER

Conclusion

____

____

Name________________________________ Date________________

• May the Force Be With You!

Purpose

How does the position of a fulcrum on a lever affect the amount of force needed to do work?

Hypothesis

____

____

____

____

Materials

Pencils, 3
Wooden ruler
Tape
Small ball of clay
Washers or pennies, 10

Procedures

1. Tape the pencils together (one on top of another) so that they form a pyramid. This is the fulcrum.
2. Place the fulcrum under the ruler at the 15 centimeter mark.
3. Put the clay on one end of the ruler. This is the load.
4. Add washers (or pennies) one at a time to the other end of the ruler until the clay is lifted. This is the amount of work required.
5. Record your results.
6. Repeat the experiment with the fulcrum at 8cm, 4cm, and 2cm *from the load.*

Observations

Distance from Load	Amount of Work Required

Conclusion

____

____

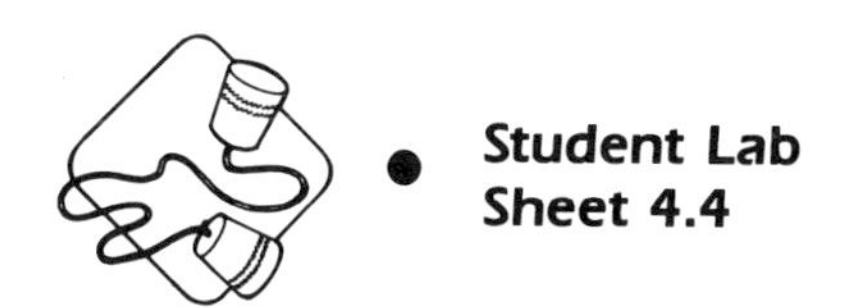

Name______________________________ Date__________________

• I Can't Hear You!

Purpose

Does sound travel better through a solid or a gas?

If you wish to make a call. . .

. . .dial your operator

Hypothesis

____

____

____

____

Materials

Bathroom-size paper cups, 2
Paper clips, 2
String, one 72-inch piece and two 24-inch pieces
Coat hanger

Procedures

1. Strike the hanger against the table. Listen to the sound.
2. Wrap the ends of the strings around the ends of each index finger. Then repeat Step 1. Put your fingers in your ears.
3. Record whether the sound was louder in Step 1 (when you heard it through the air) or in Step 2 (when you heard it through the strings near your ears).
4. Hold the paper cups apart so that the string connecting them is tight.
5. Rub the string with your thumb and index finger. Listen to the sound.
6. Repeat Step 5, but this time hold the cup next to your ear.
7. Record whether the sound was louder in Step 5 (when you heard it through the air) or in Step 6 (when you heard it through the cup next to your ear).

Observations

ACTIVITY	WHICH WAS LOUDER
Hanger sound through air	
Hanger sound through strings	
String sound through air	
String sound through cup	

Conclusion

____

____